Pile foundation for offshore wind power: macro and micro mechanism of load bearing performance

Junwei Liu, Liang Cui, Bo Han, Lingyun Feng, Xu Sun, Bo Liu

中国建筑工业出版社

图书在版编目（CIP）数据

海上风电桩基承载性能宏细观机制 = Pile foundation for offshore wind power: macro and micro mechanism of load bearing performance / 刘俊伟等著. — 北京：中国建筑工业出版社，2024.5
ISBN 978-7-112-29885-3

Ⅰ. ①海… Ⅱ. ①刘… Ⅲ. ①海上–风力发电机–发电机组–桩基础–桩承载力–研究 Ⅳ. ①TU753.6

中国国家版本馆 CIP 数据核字（2024）第 101831 号

This book consolidates the extensive research achievements of the authors and their team in the field of offshore wind turbine pile foundations.

The book consists of six chapters: Chapter 1: The current advancements in the bearing capacity of offshore wind turbine pile foundations. Chapter 2: The mechanical behavior of typical marine soils under cyclic loading. Chapter 3: The penetration dynamics of offshore wind turbine open-ended pipe piles. Chapter 4: The load-bearing characteristics of offshore wind turbine monopile foundations. Chapter 5: The cyclic shear properties at the pile-soil interface. Chapter 6: The load-bearing behavior of jacket pile foundations in offshore wind turbines.

This book presents the progress and future prospects in offshore wind turbine pile foundation research, offering significant contributions to the design and construction methodologies in offshore wind energy. It serves as an essential reference for professionals involved in the engineering and construction of offshore wind turbine foundations.

责任编辑：刘瑞霞　毕凤鸣　梁瀛元
责任校对：赵　力

Pile foundation for offshore wind power: macro and micro mechanism of load bearing performance
Junwei Liu, Liang Cui, Bo Han, Lingyun Feng, Xu Sun, Bo Liu

*

中国建筑工业出版社出版、发行（北京海淀三里河路9号）
各地新华书店、建筑书店经销
国排高科（北京）信息技术有限公司制版
北京盛通印刷股份有限公司印刷

*

开本：787毫米×1092毫米 1/16 印张：13¾ 字数：295千字
2024年5月第一版　2024年5月第一次印刷
定价：68.00元
ISBN 978-7-112-29885-3
（42899）

版权所有　翻印必究
如有内容及印装质量问题，请联系本社读者服务中心退换
电话：（010）58337283　　QQ：2885381756
（地址：北京海淀三里河路9号中国建筑工业出版社604室　邮政编码：100037）

PREFACE

Renewable energy is a strategic imperative for addressing global energy shortages and reducing carbon emissions, with wind power emerging as the most rapidly expanding, technologically advanced, and promising clean energy source. Offshore wind power, compared to its onshore counterpart, offers advantages such as higher wind speeds, greater energy yields, enhanced stability, minimal environmental impact, and no land use conflicts, making it the focal point of global wind energy development. China, endowed with extensive marine resources, possesses unique advantages for offshore wind power development.

The foundation of offshore wind turbines is a pivotal component in wind farm construction, directly influencing the safety and economic viability of the entire project. Currently, monopile foundations, which account for over 65% of offshore wind turbine foundations, are favored due to their simplicity, mature manufacturing processes, ease of transport, and straightforward construction, making them suitable for waters up to 30 meters deep. For depths between 30 and 60-70 meters, jacket pile foundations are more economical. The complex marine environment, characterized by the interplay of wind, waves, and currents, necessitates innovative solutions to reduce construction costs, enhance performance, and ensure the longevity of offshore wind turbine foundations.

This book, grounded in the extensive research and practical experience of the author and their team, systematically reviews advancements and findings in offshore wind turbine pile foundation research. It is structured into six chapters: Chapter 1: An overview of the current research on the bearing capacity of offshore wind turbine pile foundations. Chapter 2: The mechanical behavior of sandy and silty clay soils under cyclic loading. Chapter 3: The penetration dynamics of offshore wind turbine open-ended pipe piles. Chapter 4: The dynamic response of monopile foundations under complex loading conditions. Chapter 5: The cyclic shear properties at the pile-soil interface. Chapter 6: The load-bearing behavior of jacket pile foundations in offshore wind turbines.

Contributors to this volume include: Liu Junwei and Feng Lingyun from Qingdao University of Technology (Chapters 3, Section 4.2, Chapter 5, Section 6.1). Cui Liang from the University of Surrey (Section 2.1, Section 4.3). Han Bo from Shandong University (Section 2.2, Section 4.1). Sun Xu from Shandong Electric Power Engineering Consulting Institute CORP. LTD., and Liu Bo from China Power Engineering Consulting Group CO. LTD. (Chapter 1, Section 6.2). The compilation was overseen by Liu Junwei.

We extend our sincere gratitude to all individuals involved in the writing, reviewing, and support of this book.

Despite the rapid advancements in offshore wind turbine pile foundation applications, there may be oversights or inaccuracies due to the authors' limitations. Reader feedback and corrections are welcomed.

Authors
June 2024

CONTENTS

Chapter 1   Research status for bearing characteristics of pile foundation for offshore wind power 1
   1.1   Research status of typical marine soil mechanical properties under cyclic loading 2
   1.2   Research status of penetration characteristics of open-ended pipe piles 3
   1.3   Research status of bearing characteristics of monopile 6
   1.4   Research on cyclic shear characteristics of pile-soil interface...................................... 8
   1.5   Research status of bearing characteristics of pile group 10
   References 12

Chapter 2   Typical mechanical properties of marine soil under cyclic loading 16
   2.1   Responses of granular soils under cyclic loading 16
      2.1.1   Responses of granular soils in cyclic simple shear conditions 17
      2.1.2   Responses of granular soils in cyclic triaxial test conditions........................ 34
      2.1.3   Summary.. 47
   2.2   Cyclic behavior of China Laizhou Bay submarine mucky clay at an offshore wind turbine site 49
      2.2.1   Experiment process.. 50
      2.2.2   Cyclic behavior.. 55
      2.2.3   Summary.. 63
   References 64

Chapter 3   Driving characteristics of open-ended pile for offshore wind power 69
   3.1   Large scale model test on driving characteristics of open-ended pile........................ 69
      3.1.1   Test equipment and material selection... 69
      3.1.2   Test results and analyses 72
      3.1.3   Summary.. 79
   3.2   DEM investigation of installation responses of jacked open-ended piles.................. 80
      3.2.1   DEM modeling 82
      3.2.2   DEM simulation results 85

  3.2.3 Summary 97

 References 98

Chapter 4 Bearing characteristics of monopile foundations for offshore wind turbine 103

 4.1 Failure modes of rigid monopile foundation in soft clay under multidirectional loads 103

  4.1.1 Finite element numerical modeling 103

  4.1.2 Failure mode of monopile foundation under unidirectional loads 106

  4.1.3 Failure mode of monopile foundation under complex loads 110

  4.1.4 Effect of pile diameter on bearing capacity of monopile foundation............. 116

  4.1.5 Summary 117

 4.2 Dynamic response of offshore open-ended pile in sand under lateral cyclic loadings 119

  4.2.1 Studies using large-scale model tests 119

  4.2.2 Numerical studies using Discrete Element simulation 128

 4.3 Response of close-ended pile under lateral cyclic loading and the associated micro-mechanics 138

  4.3.1 Summary of scaled model tests findings and field evidence 138

  4.3.2 Numerical findings using discrete element method modeling 139

  4.3.3 Summary 148

 References 149

Chapter 5 Investigation of cyclic pile-sand interface weakening mechanism 156

 5.1 Large-scale CNS cyclic direct shear tests 157

  5.1.1 Testing apparatus 157

  5.1.2 Testing materials and testing scheme 159

 5.2 Monotonic pile-sand interface behavior 161

 5.3 Cyclic pile-sand interface weakening mechanism 161

  5.3.1 Cyclic attenuation of interface shear and normal stress amplitudes 162

  5.3.2 Stress path behavior 166

  5.3.3 Interface particle crushing 168

 5.4 Summary 170

 References 171

Chapter 6   Bearing characteristics of wind turbine jacket group pile foundation in sandy soil 174
- 6.1 Model tests of jacking installation and lateral loading performance of jacket foundation in sand 174
  - 6.1.1 Model testing 176
  - 6.1.2 Test results and discussions 180
  - 6.1.3 Summary 195
- 6.2 Dynamic response of open-ended pipe pile under vertical cyclic loading in sand and clay 196
  - 6.2.1 Materials and Methods 197
  - 6.2.2 Experimental Procedure 199
  - 6.2.3 Results and Discussions 200
  - 6.2.4 Summary 209
- References 209

Chapter 1 Research status for bearing characteristics of pile foundation for offshore wind power

Renewable energy is a strategic choice to solve the problem of global energy shortage and reduce carbon dioxide emissions, and wind power is currently the fastest growing, the most mature, and the best industrial prospect technology in multitudinous clean energies. Compared with onshore wind power, offshore wind power has many unique advantages, i.e., the high wind speed, high wind energy production, good stability, small negative impact on environment, and no occupation of land resources, which has become the main battlefield of global wind energy development.

The development of wind power resources in the world is mainly concentrated in Europe. As early as 1991, Denmark built the world's first offshore wind farm off the northwest coast of the Baltic Island of Lorraine. Since the release of the "Thirteenth Five-Year Plan for Wind Power Development" in November 2016, China's offshore wind power project construction enters the stage of comprehensive acceleration. As of the end of September 2019, the cumulative grid-connected capacity of offshore wind power in China exceeded 5 million kW, ranking third in the world after the UK and Germany. China has 18000km of coastline and can use the area of more than 3 million square kilometers. According to the "China Wind and Power Development Road Figure 2050" issued by the Institute of Energy Research and Reform Commission, within the waters with the water depth no more than 50m, China's near-sea wind energy resources can be developed up to 500 million kW, and the development of offshore wind power has excellent geographical advantages. In February 2020, Rethink Energy released a report entitled "Offshore wind power creates 8 million jobs and becomes the key to decarbonization", which predicts that by 2030, the global installed capacity of offshore wind power will increase from the current 25GW to 164GW. China will surpass the UK in 2026 and become the country with the largest installed capacity of offshore wind power in the world. By then, China's offshore wind power capacity will account for more than half of the total amount in the Asia-Pacific region, accounting for nearly a quarter of the world.

Offshore wind turbine foundation is an important part of offshore wind turbines, and it is also the key point and difficult point of wind turbine design and construction. Compared with the land environment, the marine environment is worse, the construction is more difficult, and the technical requirements are higher. The costs of onshore wind turbine foundation account for about 10% of the total costs, while the costs of offshore wind turbine foundation can reach 20% to 30%. At present, the construction of offshore wind farms is developing towards larger scale and deeper waters, which greatly avoids the near-shore environmental noise and visual pollution, as well as shipping, fishery and other sea conflicts, but also brings many problems, especially in the deep-sea conditions, turbine foundation selection and design are directly related to the construction costs of wind farms and technical difficulties. Therefore, how to reduce the costs of offshore wind turbine infrastructure construction, improve service performance and extend the duration of safety service are urgent problems to be solved in the construction of offshore wind turbines, and they are also the keys to promote the future development of offshore wind energy in China.

1.1 Research status of typical marine soil mechanical properties under cyclic loading

The bearing capacity of seabed rock and soil under cyclic loading is the key and important prerequisite of offshore wind power foundation. Studying the mechanical properties of typical marine rock and soil under offshore wind farms is of great significance for optimizing the design of offshore wind power foundations. Regarding the dynamic characteristics of marine soil, the main research methods include cyclic simple shear test, cyclic triaxial test and soil element numerical simulations. The large monopiles for offshore wind power are mainly rigid piles. Under the ultimate lateral load, the monopile foundation will be damaged by rotating around a point near the pile end, and the soil around the pile tends to be damaged by shear under the rotating effect of the pile foundation. Therefore, researchers have carried out a large number of cyclic simple shear tests to explore the dynamic characteristics of marine soils (Nikitas et al., 2017). In the study of dynamic characteristics of sandy marine soil, some researchers (Cui et al., 2019) carried out simulations of cyclic simple shear test and studied the variation law of shear modulus of sandy soil under cyclic loading and its correlation with its influencing factors. The simulation results show that the shear modulus of the sample is proportional to the density of the sample under the same conditions, and its relationship with the strain amplitude is just the opposite. This rule is also found in the model test of offshore wind power foundation (Lombardi et al., 2015) and field test (Kühn, 2000).

Compared with cyclic simple shear test, cyclic triaxial test can restore the stress environment of soil around pile foundation under dynamic loading more accurately. Therefore, this method has been widely used in the study of dynamic characteristics of marine soil. In recent years, more and more researchers have started to carry out dynamic triaxial test to investigate the dynamic characteristics of marine soil. Some researchers (Liu, 2010; Zeng and Chen, 2009; Lume et al., 2006; Zhang et al., 2018; Hu et al., 2018; Wang et al., 2013) studied the development process of stress-strain and effective stress paths in marine soil and tested the strength reduction and deformation development law of soil caused by static shear process or static shear after cyclic load through indoor dynamic triaxial test. The real-time change of stress state during the application of cyclic load needs to be further explored. At present, Numerous researchers have done loads of useful research on the law of cumulative strain development of marine soil and proposed a variety of prediction models that conform to the law of strain development. Cheng and Hou (2015) studied the dynamic strength and dynamic deformation of Zhoushan undisturbed marine clay. The cumulative plastic strain equation proposed by Monismith was used to simulate the double-amplitude strain in the experiment. The results show that the model agrees well with the experimental results. Wang and Cai (2008) studied the plastic strain development law of Hangzhou saturated clay and obtained the cumulative plastic strain-softening model by normalizing the test data. Mose et al. (2003) studied the cumulative strain of marine soils on the east coast of India. However, the existing models are often only for the single applicable models for the specimen in the damaged or undamaged state, which cannot better reflect the difference between the two strain development laws. The variation of pore water pressure of marine soil under cyclic loading is another research focus in cyclic triaxial test. Lei (2003) studied the pore water pressure characteristics of soft clay in coastal areas under cyclic loading and proposed a pore water pressure development model. Luan et al. (2011) investigated the development law of pore water pressure of undisturbed mucky soft clay in Changjiang Estuary under cyclic loading and pointed out that the development law of pore water pressure satisfies hyperbolic model. In addition, some researchers have carried out relevant research on the stiffness attenuation law of marine soils in different regions under cyclic loading. Zhu and Zheng (2014) studied the attenuation law of shear modulus of soft soil in Pukou District of Nanjing under cyclic loading and summarized the applicable shear modulus formula. Guo et al. (2018) studied the stiffness softening law of Tianjin coastal soft soil based on different load waveforms and proposed to describe the stiffness softening law with softening index. The second chapter of this book will take two typical marine soils as the research objects to explore the mechanical properties of typical marine soils under cyclic loading.

1.2 Research status of penetration characteristics of open-ended pipe piles

Another critical factor affecting the load carrying capacity of offshore wind turbine foundations is the foundation form. Currently, the common offshore wind turbine foundation types include shallow gravity-based foundation, large diameter monopile, suction bucket foundation, tripod foundation, pile group with high cap foundation, jacket foundation, etc., as shown in Figure 1.1. Among them, large diameter monopile (Figure 1.2) has the broadest range of applications. It is used in more than 80% of offshore wind farms in Europe. It is also used in a number of existing and newly built offshore wind farms in Shanghai, Jiangsu, Zhejiang, Fujian, Guangdong, and other places in China. It is made of welded steel pipe, with the wall thickness generally controlled to 1% of the pile diameter. It has the characteristics of simple structure, convenient transportation, and low construction difficulty. Its applicable water depth is generally limited to 30m. As the capacity of individual wind turbines increases and the water depth of offshore wind farms increases, higher requirements are placed on the offshore installation technology and load stability of the foundation, the jacket foundation (Figure 1.3), is also gradually becoming widely used. The concept of the jacket foundation is derived from offshore oil and gas industry. It is generally composed of the upper jacket structure, multiple corner piles or suction bucket foundations in the seabed. Its applicable water depth can reach more than 100m. In August 2009, the jacket foundation was used for the first time for six offshore wind turbines at the Alpha Ventus Wind Farm in Germany.

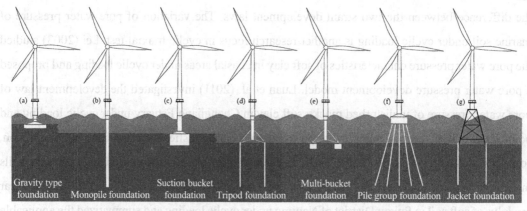

Figure 1.1   The main basic form of offshore wind turbines

Figure 1.2   Monopile foundation   Figure 1.3   Jacket foundation

Offshore wind turbine monopiles or jacket foundations mostly use open-ended pipe/tubular piles. When the pipe pile penetrates into the seabed, part of the soil is squeezed into the pile tube, forming a "soil plug-pipe pile-pile side soil" system. The dynamic development process of soil plugs is complex; even for large-diameter steel pipe piles, the height of the earth plug and the depth of the driven pile are not synchronized, and the final stage of pile driving is prone to show blocking, hammer rejection, and other engineering problems. For layered soils, plugging and opening may undergo multiple alternating stages, depending on the magnigude of the resistance of the pile end soil and the resistance of the inner wall. Especially for complex seabeds with substantially spatial intensity variability, the plugging effect is more complicated. The height of the soil plug is different under different working conditions, and there are differences in the pile-soil stiffness and interfacial frictional characteristics. Research on the formation and vertical action mechanism of soil plugs in a single soil layer is becoming increasingly abundant. The soil plug effect for layered soils has gradually attracted attention all over the world. However, under complex marine geological conditions, the formation of soil plugs and their influence on the bearing characteristics of large-diameter pipe piles are still inconclusive, especially because the mechanism of actions of horizontal bearing characteristics has not attracted enough attentions. In addition, there is also a weakening effect of the soil plug itself under cyclic load. The weakening effect of soil plug on the dynamic response of large-diameter pipe piles needs further investigations. Therefore, clarifying the soil plug formation and load transfer mechanism are crucial for the piling characteristics of offshore wind turbine pile foundations and the subsequent horizontal bearing capacity.

In the process of driving the pile, part of the soil is squeezed into the pipe pile to form a "soil plug". The soil plug effect is the primary reflection of the differences between the open-ended pipe pile and the close pipe pile or the solid pile. It is also the main reason for the differences in the bearing characteristics of the two (Randolph, 2003). Studying the formation and load transfer mechanism of soil plugs in the process of pile driving is the key to accurately estimating the pile performance of open-ended pipe piles. The installation of pile shoes and the change of pile diameter are found to change the formation of soil plugs, the pile sinking characteristics and bearing performance of open-ended pipe piles. The bearing capacity of an open-ended pile is composed of pile wall friction, pile end bearing capacity, and soil plug resistance. Among them, the frictional resistance is provided by the relative displacement between the pile body and the earth plug. However, the study found that the bearing mechanism of internal and external friction is different in actual scenarios. In the initial stage of the pile driving, the external friction force of the pile develops from top to bottom with the displacement of the pile body, and the full load applied to the pile top is resisted by the friction force along the outside of the pile. At this stage, no soil has been squeezed into the inside of the steel pipe pile, so there is no friction along the inside of the pile. With the further increase of the load, the load is gradually transferred to the soil at the bottom of the steel pipe pile, and the soil underneath the pile is squeezed into the pile to form a certain length of soil plug, which maintains the state of the soil plug during the operation of the pile foundation (Paikowsky et al.,1990). To reveal the load transfer mechanism of pile foundation, some researchers (Gavin and Lehane, 2007; Randolph and Wroth, 1978; Randolph, 2003) have carried out many relevant studies and proposed some theoretical models of load transfer. However, these models are primarily based on experiments on solid piles, and there is a certain degree of inadequacy for open-ended pipe piles. Recently, some researchers (Yu and Yang, 2012; White and et al., 2000; Lehane and Gavin, 2001) have carried out indoor and field experiments on the plug effect, which has dramatically improved the understanding of the plug effect of open-ended pipe piles. However, the available model analysis of pipe piles is still relatively lacking, which limits the application of theoretical analysis and prediction methods for the bearing capacity and settlement of open-ended piles to a certain extent. The third chapter of this book will use the extensive scale model tests and discrete element simulations to study the driving characteristics of open-ended pipe piles.

1.3 Research status of bearing characteristics of monopile

Offshore wind turbine foundation is subject to coupled horizontal loads such as ocean wind, wave and current loads. Compared with onshore wind turbine foundation, offshore wind turbine foundation has a harsher service environment, more complex failure mechanism and difficult design. For offshore wind turbine foundations in service, the load forms are very complex. Unlike offshore oil and gas platforms, offshore wind turbine foundations are subjected to self-weight (V) from the upper structure, horizontal loads (H) of wind, wave and current transmitted by blades and tower during service, and large overturning bending moment (M) generated by horizontal loads, as shown in Figure 1.4. Therefore, bearing capacity of wind turbine foundation under multi-directional load is an important issue in design. For monopile foundation, the bearing capacity design mainly considers the overturning moment caused by wind wave. At this time, the monopile foundation is in the "V-H-M" three-way composite stress state. For four-pile jacket foundation, when the overturning bending moment is large, each pile will change from V-H-M composite bearing capacity to tension-compression mode, i.e. the overturning bending moment transmitted by the upper structure is resisted mainly by the uplift and pushdown force of the front and rear foundation piles in the loading direction.

Chapter 1 Research status for bearing characteristics of pile foundation for offshore wind power

Figure 1.4 Loading diagram of offshore wind turbine foundation

As mentioned above, monopile foundation is the simplest type of offshore wind turbine foundation. Its design and construction technology is relatively simple, and it is easy to popularize, apply and industrialize. It is the most widely used foundation type for offshore wind power plants all over the world (Esteban and Leary, 2012). However, there is no uniform understanding about the bearing characteristics and failure modes of monopile foundation under coupled complex load. Monopile foundation of offshore wind turbine bears complex long-term environmental loads such as wind, wave and current at the same time, which results in foundation damage and seriously affects its normal operation (Luengo et al., 2019). Therefore, researchers have carried out a lot of research on damage mode of monopile foundation under complex marine loads. The existing failure modes of pile foundation under complex loads mainly focus on the ultimate bearing capacity under different loads and the interaction mechanism between pile and soil. For the ultimate bearing capacity of pile foundation, researchers (Matsumoto et al., 2004; Han et al., 2015) proposed various methods to determine the ultimate bearing capacity of pile foundation under complex loads through physical model test and numerical simulations. Sastry and Meyerhof (1999) studied the stress distribution of soil around piles under ultimate lateral load by conducting homogeneous soil model tests in laboratory, and deduced the empirical formula of ultimate bearing capacity of homogeneous soil layer under lateral load. On this basis, some researchers (Liu et al., 2014; Zormpa and Comodromos, 2018) no longer limited to the study of ultimate bearing capacity under a single load, but explored the interaction of different loads and their influence on bearing capacity. Zhukov and Balov (1978) carried out field tests on bearing capacity of monopile in clay. The analysis results show that the vertical load at the top of monopile can improve the horizontal bearing capacity of monopile to a certain extent. However, this conclusion led to enthusiastic

7

discussions in academic and engineering fields. A group of researchers represented by Karthigeyan et al. (2007) attempted to discuss it from a theoretical point of view, and their analytical calculation results were completely opposite to those of Zhukov and Balov. A large number of aforementioned research results indicate that the loading sequence of vertical load and horizontal load on the top of pile has a great influence on the bearing capacity of pile foundation. However, there is no uniform understanding of the influence of various load coupling modes on the bearing capacity of pile foundation under complex loads. In the fourth chapter of this book, the bearing characteristics of monopile under complex load coupling are studied based on numerical simulations, model tests and field tests.

1.4 Research on cyclic shear characteristics of pile-soil interface

The long-term service performance issues of offshore wind turbine foundation are prominent. Wind, waves, currents and other marine loads have been in the service of offshore wind turbines for a long time, including waves and wind cycle numbers up to 10^7-10^8. Due to the weakening of the soil strength at soil-foundation contact surface under long-term cyclic loading, the foundation may undergo cyclic cumulative deformation and load-bearing capacity changes, thus affect the normal operation of wind turbines and even lead to the failure of the entire structure. The design specifications of wind turbines in various countries require strict control of the cyclic cumulative angle of the wind turbine foundation during the service life. For example, the German specifications stipulate that the total angle of foundation rotation under installation and long-term operation shall not exceed 0.5°; the maximum permissible cyclic cumulative angle of foundation of Thornton Bank offshore wind farm in the UK is 0.25°; and China's "Design Regulations for Wind Farm Foundation" promulgated in 2007 (FD003-2007) stipulates that the allowable tilt angle of large wind turbine towers with a hub height greater than 100m is only 0.17° during the service life. In addition, wind power support structures are dynamic-sensitive systems with naturual (self-oscillation) frequencies very close to the turbine frequency (1P) and blade crossing frequency (2P/3P), as well as conventional wind and wave loads, as shown in the statistics in the UK wind power engineering manual (Figure 1.5). The foundation is the support of the upper load and is also the dominant factor in determining the natural frequency. Therefore, when designing the foundation for offshore wind power, it is important to strictly ensure that the natural frequency avoids the environmental load frequency, and the "soft-stiff" design principle is usually used to keep the foundation natural frequency between the 1P and 3P frequency bands.

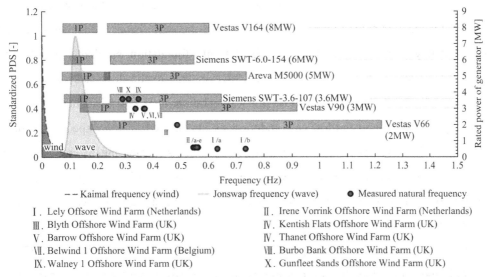

Figure 1.5 Frequency diagram of offshore wind turbine

However, under the long-term cyclic loading combined with the dynamic load generated by the rotation of the turbine rotor and the dynamic shadow load generated by the passage of the blade, the foundation-seabed system evolves gradually, e.g., the pile-soil contact stiffness changes continuously, and the natural frequency of the support system changes, which may lead to resonance of the structure and accelerated fatigue damage. For example, the first-order natural frequency of the wind turbine at the Lely wind farm in the Netherlands increased from 0.4Hz to the 1P frequency band (center frequency of 0.53Hz) after six years of operation, resulting in a significant reduction in the service life of the wind turbine. Therefore, the cyclic shear characteristics of the pile-soil interface are a key issue for the long-term service performance of wind turbine foundations.

Wind turbine foundations are in long-term service under cyclic loads such as wind and waves, and the reduction of pile-soil contact surface stiffness is likely to cause sliding of the pile-soil interface and result in loss of bearing capacity (Randolph, 2003). However, due to the key dynamic and geotechnical factors such as soil dynamic properties, interface stress conditions, and pile type under designed cyclic loads, the current understanding of the pile-soil interface weakening mechanism is not thorough enough. For this reason, some researchers have studied the pile-soil interface behavior under cyclic shear loading in order to better understand the pile-soil interface weakening mechanism and to explain the principle of pile-soil interface conditions inducing pile failure under cyclic loading. Some researchers have conducted studies on the behavior of pile-soil interface under cyclic loading by adopting the methods of in-situ tests (Chow, 1997; Tsuha et al., 2012; Buckley et al., 2018) and indoor model tests (Rimoy, 2013; Li et al., 2012; Zhou et al., 2019).

Full-scale in-situ tests can represent the real stress state of the soil, and the test results are not affected by scaling effects or sample perturbations. On the other hand, indoor tests can define various boundary conditions to simulate different pile foundation problems, and the result interpretation process is more straightforward. However, due to the drawback of higher costs, full-scale field tests are usually only used for pile testing before installation in actual projects. With the advantages of short time and low cost, indoor tests can be used for more specific quantitative study of the interface properties of pile-soil interaction problems considering different geotechnical materials, pile types, and loading conditions. Some researchers realized the limitations of field tests and carried out a large number of indoor tests. However, the research results are still very limited, and these works are focused on the pile-soil interface properties under single load, and there is a huge lack of research on the pile-soil interface properties under cyclic load. In Chapter 5 of this book, the study of cyclic shear properties of the pile-soil interface will be carried out to address the shortage of existing studies.

1.5 Research status of bearing characteristics of pile group

Although the monopile foundation is currently the most widely used type of foundation for offshore wind farms, it has an inherent limitation that it is only suitable for shallow water areas. Monopile foundation is generally used in sea areas with water depth up to 30-40m; for sea areas with water depths exceeding this depth but less than 60-70m, jacket foundation is generally used, such as three-pile and four-pile jacket foundation; for sea areas with water depths reaching or even exceeding 80-100m, the use of fixed foundation can no longer meet the requirements of economy or safety. At this time, floating foundation is a better choice. At present, with the exhaustion of offshore resources, wind power developers are beginning to advance to deep seas. Excessively deep sea water will increase investment and construction costs, so it is very important to select the appropriate foundation form of wind turbines in deeper sea areas. At present, the industry regards 50m water depth as a "watershed", but Xodus, an influential consulting agency in the offshore wind power industry, published a report based on relevant studies, overturning the existing cognition of the industry. The report believes that in the water depth range of 90m, it is more economical to use super-large jacket foundation for offshore wind turbines than floating platform foundation. Compared with monopile foundation, offshore wind turbine jacket foundation can adapt to deeper water depths, greatly expanding the sea areas where wind resources can be exploited. Compared with the floating foundation which is still in the test stage, the technology of jacket foundation is more mature; the cost is lower; and it can be popularized and applied more quickly in a short time. It can be predicted that offshore wind power will usher in the era of "jacket foundation".

At present, many researchers have studied and explored pile group foundations such as jacket foundation. Singh and Prakash (1971) studied the dynamic response of pile group in sand under lateral cyclic loading, and concluded that the turning angle and horizontal displacement of pile group cap would increase with the increase of cycle number, and the vertical bearing capacity had little effect on it. Brown et al. (1987) carried out a field test on the force-deformation relationship of pile group foundation in hard clay under bidirectional cyclic loading. The results show that there is a strong nonlinearity between pile and soil under bidirectional cyclic loading, and the final soil resistance attenuation of pile group foundation is more obvious than that of monopile foundation. The load of the front row of piles in pile group is greater than that of the back row of piles, and the load of the back row of piles decreases in turn. Chen and Chen (2013) discussed the pile-soil interaction under horizontal cyclic loading, and found that the load distribution of each row of piles under horizontal cyclic loading presents the rule of 'large in the front row, small in the back row'. Considering the difficulty of field test, researchers have studied the loaded characteristics of jacket foundation by numerical simulation and model test. Mostafa and El Naggar (2004) used the finite element software SACS to conduct a detailed parametric analysis of the response of jacket foundation for wind turbines under extreme wind and wave loads. Elshafey et al. (2009) studied the dynamic response of jacket foundation from the perspective of model test and numerical theory. Mirza et al. (2015) used the computer program SEADYN to numerically model the Beihai jacket platform. The pile-soil interaction considered the nonlinear spring in the API specification. It is found that the axial force and bending moment at the top of each pile foundation are different, and are obviously affected by the axial and transverse foundation reaction modulus. Li and Li (2012) used finite element software to analyze the bearing characteristics of the pile foundation of the existing jacket platform under horizontal loading. Yuan et al. (2012) used nonlinear finite element analysis method to study the interactions between pile and soil of jacket foundation under horizontal load. They analyzed the influence of the stiffness and diameter of model pile, soil parameters including horizontal earth pressure coefficient and dilation angle on the bearing characteristics of the foundation. Run (2012) used finite element software to simulate nonlinear three dimensional pile-soil interaction to study the bearing capacity of pile foundation of jacket platform under inclined loading. The study showed that vertical loading improved the horizontal bearing capacity of pile foundation to a certain extent. With the increase of vertical loading, the lateral displacement of pile top has a decreasing trend. This conclusion was also confirmed by physical model test. The above investigations on jacket foundation use numerical analysis and normal gravity scale model test. For the normal gravity scale model test, the soil stress around the model structure is not accurate, and it is difficult to truly reflect the catastrophe collapse of the actual rock-soil mass. For numerical analysis, the validity of the results needs to be further verified. In short, the research on the stress and deformation characteristics of jacket foundation under horizontal static and cyclic loading is relatively limited. The Chapter 6 of this book will explore the bearing characteristics of jacket pile group foundation for offshore wind power on the basis of existing studies.

In conclusion, the dynamic responses characteristics of the foundation and seabed soil system under the actions of complex marine environmental loads are the key factors that determine the service life of the offshore wind power support system. In view of the deficiencies of existing studies, this book presents relevant studies from four aspects: typical characteristics of marine soil mechanics, driving characteristics of open-end pipe pile, bearing characteristics of monopile, and cyclic shear behavior of pile-soil interface. It provides theoretical support and technical guidance for offshore wind turbine foundation design and optimization. Finally, this book describes prospective researches on the front field of offshore wind power development, explores the bearing characteristics of pile group supporting offshore wind turbine, and provides theoretical support for the development of offshore wind power to the far and deep sea.

References

Brown D A, Reese L C, O'Neill M W. Cyclic lateral loading of a large-scale pile group. Journal of Geotechnical Engineering, 1987, 113(11): 1326-1343.

Buckley R M, Jardine R J, Parker D, et al. Ageing and cyclic behaviour of axially loaded piles driven in chalk. Geotechnique, 2018, 68(2): 146-161.

Chen S, Chen F. Research on the behavior of pile groups under cyclic lateral loads. Soil Engineering and Foundation, 2013, 27(4): 100-103.

Cheng Y H, Hou H. Test study on dynamic strength and strain behavior of representative marine soil. Soil Engineering and Foundation, 2015(3): 161-165.

Chow F C. Investigations into displacement pile behaviour for offshore foundations. Ph D. Thesis. Imperial College, University of London, 1997.

Cui L, Bhattacharya S, Nikitas G, et al. Micromechanics of granular soil in asymmetric cyclic loadings: an application to offshore wind turbine foundations. Granular Matter, 2019, 21.

Elshafey A, Haddara M R, Marzouk H. Dynamic response of offshore jacket structures under random loads. Marine Structures, 2009, 22(3): 504-521.

Esteban M, Leary D. Current developments and future prospects of offshore wind and ocean energy. Applied Energy, 2012, 90(1): 128-136.

Gavin K, Lehane B. Base load-displacement response of piles in sand. Canadian Geotechnical Journal, 2007, 44(9): 1053-1063.

Guo F, Cheng J, Li Z. Effect of cyclic loading waveforms on stiffness softening characteristics of soft soil in Tianjin Binhai New Area. Port & Waterway Engineering, 2018(7): 148-154.

Han B, Zdravkovi L, Kontoe S. The stability investigation of the generalised-a time integration method for dynamic coupled consolidation analysis. Computers and Geotechnics, 2015, 64: 83-95.

Hu X, Zhang Y, Guo L, et al. Cyclic behavior of saturated soft clay under stress path with bidirectional shear stresses. Soil Dynamics and Earthquake Engineering, 2018, 104: 319-328.

Karthigeyan S, Ramakrishna V, Rajagopal K. Numerical investigation of the effect of vertical load on the lateral response of piles. Journal of Geotechnical and Geoenvironmental Engineering, 2007, 133(5): 512-521.

Kühn M. Dynamics of offshore wind energy converters on monopile foundations-experience from the Lely offshore wind farm. OWEN Workshop "Structure and Foundations Design of Offshore Wind Turbines" March 1, 2000, Rutherford Appleton Lab.

Lehane B M, Gavin K. Base resistance of jacked pipe piles in sand. Journal of Geotechnical & Geoenvironmental Engineering, 2001, 127(6): 473-480.

Lei H Y. Pore pressure model of marine soft soil and its parameters in Tianjin region. Journal of Jilin University Earth Science edition, 2003(1): 76-79.

Li G L, Li C J. Analysis of the horizontal bearing capacity of the jacket platform single pile in saturated clay. China Petroleum Machinery, 2012, 40(10): 63-66.

Li Z, Bolton M D, Haigh S K. Cyclic axial behaviour of piles and pile groups in sand. Canadian Geotechnical Journal, 2012, 49(9): 1074-1087.

Liu G X. Study on shear behaviors of saturated marine soils under complex stress conditions. Dalian: Dalian University of Technology, 2010.

Liu J W, Guo Z, Zhu N, et al. Dynamic response offshore open-ended pile under lateral cyclic loadings. Journal of Marine Science and Engineering, 2019, 7(5).

Liu M, Yang M, Wang H. Bearing behavior of wide-shallow bucket foundation for offshore wind turbines in drained silty sand. Ocean Engineering, 2014, 82(15): 169-179.

Lombardi D, Bhattacharya S, Scarpa F, et al. Dynamic response of a geotechnical rigid model container with absorbing boundaries. Soil Dynamics and Earthquake Engineering, 2015, 69: 46-56.

Luan M T, Liu G X, Guo Y, et al. Characteristics of pore water pressure and strength of undisturbed saturated marine clay under complex stress conditions. Chinese Journal of Geotechnical Engineering, 2011(1): 150-158.

Luengo J, Negro V, Garcia-Barba J, et al. New detected uncertainties in the design of foundations for offshore wind turbines. Renewable Energy, 2018, 131: 667-677.

Lunne T, Berre T, Andersen K H, et al. Effects of sample disturbance and consolidation procedures on measured shear strength of soft marine Norwegian clays. Canadian Geotechnical Journal, 2006, 43(7): 726-750.

Matsumoto T, Fukumura K, Kitiyodom P, et al. Experimental and analytical study on behaviour of model piled rafts in sand subjected to horizontal and moment loading. International Journal of Physical Modelling in Geotechnics, 2004, 4(3): 1-19.

Moses G, Rao S N, Rao P N. Undrained strength behaviour of a cemented marine clay under monotonic and cyclic loading, 2003, 30(14): 1765-1789.

Mostafa Y E, El Naggar M H. Response of fixed offshore platforms to wave and current loading including soil-structure interaction. Soil Dynamics and Earthquake Engineering, 2004, 24(4): 357-368.

Nikitas G, Arany L, Aingaran S, et al. Predicting long term performance of offshore wind turbines using cyclic simple shear apparatus. Soil Dynamics and Earthquake Engineering, 2017, 92: 678-683.

Paikowsky S G, Whitman R V, Baligh M. A new look at the phenomenon of offshore pile plugging. Marine Geotechnology, 1990, 8(3): 213-230.

Randolph M F, Leong E C, Houlsby G T. One dimensional analysis of soil plugs in pipe piles. Geotechnique, 1991, 41: 587-598.

Randolph M F, Wroth C P. Analysis of deformation of vertically loaded piles. J. Geotech. Eng. Div. 1978, 104: 1465-1488.

Randolph M F. RATZ Program Manual: Load transfer analysis of axially loaded piles; Dept. of Civil and Resource Engineering, University of Western Australia: Perth, Australia, 2003.

Randolph M F. Science and empiricism in pile foundation design. Géotechnique, 2003, 53: 847-875.

Rimoy S P. Ageing and axial cyclic loading studies of displacement piles in sands. Ph D. Thesis. Imperial College, University of London, 2013.

Run J H. The bearing capacity analysis of pile foundation of jacket platform under inclined load. Shanghai: Shanghai Jiao Tong University, 2012.

Sastry V R N, Meyerhof G. Flexible piles in layered soil under eccentric and inclined loads. Soils and Foundations, 1999, 39(1): 11-20.

Singh A, Prakash S. Model pile group subjected to cyclic lateral load. Soils and Foundations, 1971, 11(2): 51-60.

Tsuha C, Foray P Y, Jardine R J, et al. Behaviour of displacement piles in sand under cyclic axial loading. Soils and Foundations, 2012, 52(3): 393-410.

Wang J, Cai Y Q. Study on accumulation plastic strain model of soft clay under cyclic loading. Chinese Journal of Rock Mechanics and Engineering, 2008(2): 331-338.

Wang J, Guo L, Cai Y, et al. Strain and pore pressure development on soft marine clay in triaxial tests with a large number of cycles. Ocean Engineering, 2013, 74: 125-132.

White D J, Sidhu H K, Finlay T C R, et al. Press-in piling: The influence of plugging on drivability. In Proceedings of the 8th International Conference of the Deep Foundations Institute, New York, NY, USA, 5-7 October, 2000, 299-310.

Yu F, Yang J. Base capacity of open-ended steel pipe piles in sand. Journal of Geotechnical & Geoenvironmental Engineering, 2012, 138(9): 1116-1128.

Yuan Z L, Duan M L, Chen X Y, et al. Nonlinear finite element analysis of jacket platform pile foundations under lateral loads. Rock and soil mechanics, 2012, 33(8): 2551-2560.

Zeng L, Chen X P. Analysis of mechanical characteristics of soft soil under different stress paths. Rock and soil mechanics, 2009(5): 1261-1270.

Zhang S, Ye G, Liao C, et al. Elasto-plastic model of structured marine caly under general loading conditions. Applied Ocean Research, 2018, 76: 211-220.

Zhou W J, Wang L Z, Guo Z, et al. A novel t-z model to predict the pile responses under axial cyclic loadings. Computers and Geotechnics, 2019, 112: 120-134.

Zhu Z R, Zheng L. Research on stiffness degradation of saturated clay under cyclic loading. Journal of Hydraulic and Architectural Engineering, 2014(5): 180-184.

Zhukov N V, Balov I L. Investigation of the effect of a vertical surcharge on horizontal displacement and resistance of pile columns to horizontal load. Soil Mechanics and Foundation Engineering, 1978, 15 (1): 16-22.

Zormpa T E, Comodromos E M. Numerical evaluation of pile response under combined lateral and axial loading. Geotechnical and Geological Engineering, 2017, 36: 793-811.

Chapter 2 Typical mechanical properties of marine soil under cyclic loading

To obtain detailed information on the soil-foundation interaction for offshore wind turbines under various conditions, it is better to consider a small sample of soil close to the foundation and replicate in element tests with similar stress conditions and stress paths. This chapter presents a study of the properties of marine soils around monopile and jacket foundations. Section 2.1 presents the cyclic behaviors of sandy soils, while Section 2.2 presents the cyclic behaviors of cohesive soils.

2.1 Responses of granular soils under cyclic loading

To attain detailed information of soil-foundation interactions for offshore wind turbine under various conditions, it is better to consider a small zone of soils adjacent to the foundation and replicate it in an element test with similar stress conditions and stress paths. Majority of current offshore wind turbines are supported by monopile foundation, which is a large steel tube/pipe typically 30-40m in length and 3-7m in diameter. Unlike the slender piles for offshore structure, monopile tends to rotate rather than bend under lateral load or overturning moment. Therefore, the interactions between the monopile and the soil element in front of the monopile could be represented by cyclic simple shear tests, as shown in Figure 2.1 (Cui et al., 2019). Detailed descriptions of responses of granular soils in cyclic simple shear tests were described in Section 2.1.1.

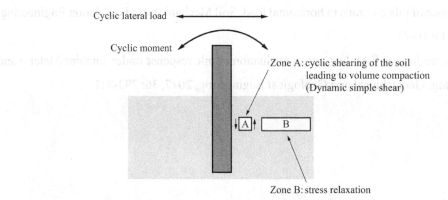

Figure 2.1 Schematic diagram of soil stress conditions surrounding a monopile (Cui et al., 2019)

Apart from the monopile foundations for shallow water depth (typically < 30m), jacket foundation was frequently used to support the offshore wind turbine with suction caissons or multi-piles fixing to the seabed. Under the lateral cyclic loading, jacket swings/tilts cyclically, leading to the cyclic push and pull of suction caissons or multi-piles. Soils under these structures are subjected to cyclic triaxial test conditions. To understand the soil structure behaviors, a detailed description of responses of granular soils in cyclic triaxial tests were described in Section 2.1.2.

2.1.1 Responses of granular soils in cyclic simple shear conditions

2.1.1.1 Experimental tests

Test material and test program

A cyclic simple shear apparatus manufactured by VJ Tech was used by Nikitas et al. (2016) for testing cylindrical soil samples. RedHill 110 Sand, a poorly graded fine grained silica sand with $d_{50} = 0.18$mm and particle size distribution (PSD) curve shown in Figure 2.2, was tested as this soil has been used to carry out scaled model tests on different types of foundations (e.g. Lombardi et al., 2015; Kelly et al., 2006). The sand has a specific gravity, G_s, of 2.65 and minimum and maximum void ratio of 0.608 and 1.035 respectively. Specimens of 50mm in diameter and 20mm in height were prepared for testing, as suggested in ASTM D6528 (2007). Strain controlled cyclic simple shear tests on medium dense sand (relative densities $D_r = 50\%$) and dense sand ($D_r = 75\%$) were performed. Tests were carried out with various vertical stresses ($\sigma = 50$kPa, 100kPa, 200kPa) and shear strain amplitudes ($\gamma_{max} = 0.1\%, 0.2\%, 0.5\%$). The various vertical stresses represent soil blocks along monopile at various depths, while various shear strain amplitudes represent different pile deflections/rotations.

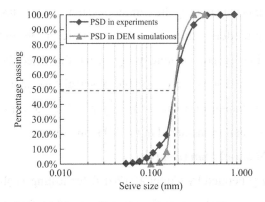

Figure 2.2 Particle size distribution (PSD) of test sand in experiments (Nikitas et al., 2016) and DEM simulations (Cui et al., 2016)

Test results and discussions

The variations of shear modulus are illustrated in Figure 2.3(a). The shear modulus increases rapidly in the initial loading cycles and then the rate of increase diminishes and the shear modulus remains below an asymptote. The shear modulus increases with increasing vertical stress and relative density, but decreases with increasing strain amplitude as expected.

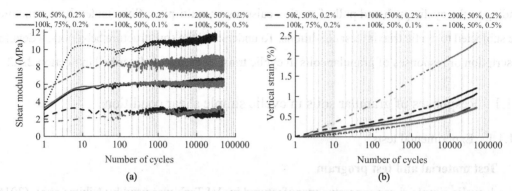

Figure 2.3 (a) Variation of shear modulus in experiments; (b) Accumulated vertical strain in experiments (Legend: σ, D_r, γ_{max}) (after Nikitas et al., 2016)

The accumulated vertical strains are illustrated in Figure 2.3(b). The increase of shear modulus is underlain by the consistent contractive responses of all samples. It can also be observed that vertical accumulated strain is proportional to the shear strain amplitude and vertical stress, but inversely proportional and relative density of soil. The results correlated quite well with the observations from scaled model tests with different types of offshore wind turbine foundations (Cuéllar et al., 2012; Bhattacharya et al., 2013a, b; Lombardi et al., 2013) and the field measurements (Kühn, 2000).

2.1.1.2 DEM simulations

A commercial DEM code PFC2D (Itasca, 2008) was used by Cui et al. (2019) to perform the DEM simulations. The experimental cyclic simple shear test is a three-dimensional problem; however, their study only models a thin slice of the sample in the middle. It is obvious that a two-dimensional simulation cannot accurately represent the three-dimensional granular soil. However, there was no intention to reproduce the physical test quantitatively but analyze the similar underlying micro-mechanism because the major and minor principal stresses in a simple shear test lie in the loading plane considered in the 2D simulations.

Description of DEM simulation program

The sample initially generated by Cui et al. (2019) for testing is about 20mm in height and 50mm in width, similar to the sample dimensions in experiment performed by Nikitas et al. (2016). It contains 8000 disks with size ranging from 0.1mm to 0.3mm and $d_{50} = 0.18$mm, matching the d_{50} value in experiments. The PSD curve is also given in Figure 2.2. Note that the PSD range is narrower in the DEM than that in experiments due to the unrealistic computational time required for simulating samples with the same particle grading as in experiments. Parameters used in DEM simulations are listed in Table 2.1. Two sets of samples were generated with radius expansion approach followed by K_0-consolidation with particle friction coefficient, $\mu = 0$ and 1.0, to generate relative dense and loose samples, respectively. Particle friction coefficient was then changed to 0.5 and specimens were brought to equilibrium again with the new frictional coefficient. Final void ratios (e) of the relative dense and loose samples at two different vertical stresses (σ) are also listed in Table 2.1. Note that due to different boundary velocity in consolidation stage, the loose sample reached slightly lower e at 50kPa than that at 100kPa.

DEM simulation parameters (Cui et al., 2019)   Table 2.1

DEM parameter	Value
Particle density	2650kg/m³
Frictional coefficient	0.5
Normal stiffness of particle	8.0×10^7 N/m
Shear stiffness of particle	4.0×10^7 N/m
Normal and shear stiffness of boundary	4.0×10^9 N/m
Vertical stress	50kPa, 100kPa
Void ratio	Dense: 0.185 ($\sigma = 50$kPa), 0.181 ($\sigma = 100$kPa) Loose: 0.215 ($\sigma = 50$kPa), 0.227 ($\sigma = 100$kPa)

Jalbi et al. (2019) reviewed the wind and wave loading data from 15 offshore wind farms and determined the maximum and minimum moment applied at the mudline of a wind turbine. It was found that ζ_c (the ratio of minimum moment to the maximum moment) lies in the range between -0.5 and 0.5 for these wind farms. This is a combined effect of wind load with high magnitude but low frequency and wave load with low magnitude but high frequency. With increasing water depth, the magnitude of wave load increases and may change the one-way cyclic loading caused by wind load into a two-way loading. A schematic diagram illustrating the variations is given in Figure 2.4(a). The cyclic loading profiles considered by Cui et al. (2019) are illustrated in Figure 2.4(b) and listed in Table 2.2, which embrace the load scenarios summaries by Jalbi et al. (2019) as well as previous studies. Similar to the parameter, ζ_c, used by Zhu et al. (2013), a strain ratio, $\eta_c = \gamma_{min}/\gamma_{max}$, was defined by Cui et al. (2019) to quantify the degree of asymmetry of cyclic loading as strain-controlled cyclic tests were performed in the current study. Strain ratios, η_c, between 0.5 and -1.0 were considered. Various soil packing densities and vertical stresses (Series A), various strain magnitudes (Series B) were considered. Twelve measurement circles were defined within the sample to measure the average stress, void ratio and coordination number. The results presented in the following sections are the average value from these twelve measurement circles.

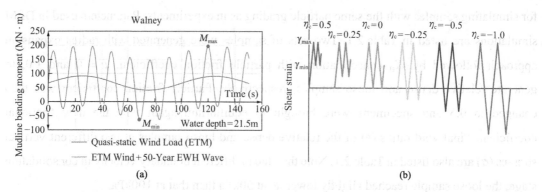

Figure 2.4 (a) Combination of wind and wave load; (b) One-way and two-way cyclic loading patterns in current DEM simulations (Cui et al., 2019)

DEM simulation program (Cui et al., 2019)　　　　　Table 2.2

Simulation series	Simulation ID	(γ_{min}, γ_{max})	σ (kPa)	Packing density
A	A-1		50	Loose
	A-2		100	Loose
	A-3*	(−0.52%, 0.52%)	50	Dense
	A-4		100	Dense
B	B-1	(−0.10%, 0.10%)		
	B-2*	(−0.52%, 0.52%)	100	Dense
	B-3	(−1.04%, 1.04%)		
C	C-1*	(−0.52%, 0.52%) ($\eta_c = -1$)		
	C-2	(−0.29%, 0.75%) ($\eta_c = -0.39$)	100	Dense
	C-3	(0, 1.04%) ($\eta_c = 0$)		
D	D-1	(−0.92%, 0.92%) ($\eta_c = -1$)		
	D-2	(−0.46%, 0.92%) ($\eta_c = -0.5$)		
	D-3	(−0.23%, 0.92%) ($\eta_c = -0.25$)	100	Dense
	D-4	(0, 0.92%) ($\eta_c = 0$)		
	D-5	(0.23%, 0.92%) ($\eta_c = 0.25$)		
	D-6	(0.46%, 0.92%) ($\eta_c = 0.5$)		
E	E-1	(−0.92%, 0.92%) ($\eta_c = -1$)		
	E-2	(−0.46%, 0.92%) ($\eta_c = -0.5$)		
	E-3	(−0.23%, 0.92%) ($\eta_c = -0.25$)	100	Loose
	E-4	(0, 0.92%) ($\eta_c = 0$)		
	E-5	(0.23%, 0.92%) ($\eta_c = 0.25$)		
	E-6	(0.46%, 0.92%) ($\eta_c = 0.5$)		

*Same simulations.

DEM simulation results at macro-scale

The shear stress-shear strain curve forms hysteresis loops during cyclic loadings. Stress-strain relationships for the symmetric two-way loading (Series A, B, C) are similar. Results from Series E were presented in Figure 2.5. While η_c decreases from 0.5 to -1.0 (one-way loading to symmetric two-way loading), stress-strain relationships also show different evolution trends. With $\eta_c = 0.5$, shear strain only reverses half way, shear stress reduces to a positive value slightly above zero in the first cycle. With the progress of cyclic loading, shear stress at both γ_{max} and γ_{min} deceases steadily. As a consequence, the negative τ_{min} at γ_{min} approaches the same magnitude of the positive τ_{max} at γ_{max}, even though both γ_{max} and γ_{min} are positive values. With decreasing η_c, the τ_{min} decreases more as γ_{min} decreases. However, the trends of τ_{max} are quite different, even though γ_{max} remains the same. When $\eta_c > 0$, τ_{max} decreases with cyclic loading; when $\eta_c = 0$, τ_{max} remains almost constant during cyclic loading; when $\eta_c < 0$, τ_{max} increases with cyclic loading. It demonstrates that the magnitude of γ_{min} affected the stress response at γ_{max}. Following 1000 number of loading cycles, no matter the symmetry of loading cycles (one-way or two-way), shear stress oscillates two-way symmetrically, i.e., negative τ_{min} at γ_{min} reaches the same magnitude as the positive τ_{max} at γ_{max}.

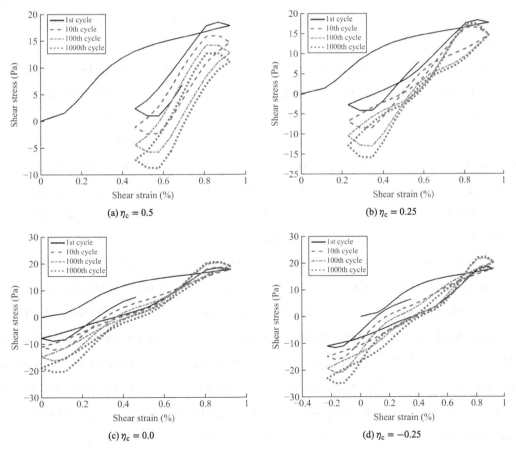

(a) $\eta_c = 0.5$ (b) $\eta_c = 0.25$

(c) $\eta_c = 0.0$ (d) $\eta_c = -0.25$

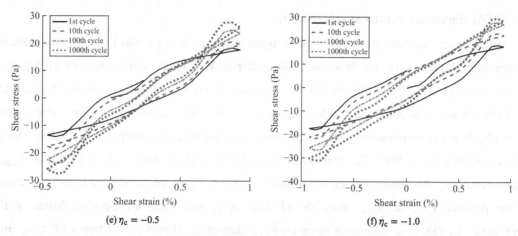

Figure 2.5 Stress-strain relationships during cyclic loadings for simulation Series E
(Cui et al., 2019)

The shear modulus (G) of the sample was calculated as

$$G = \frac{\tau_{max} - \tau_{min}}{\gamma_{max} - \gamma_{min}} \tag{2.1}$$

The variations of G in all simulations under cyclic loading are illustrated in Figure 2.6. The magnitudes of shear moduli in the DEM simulations are in the same range as the values from the experimental tests in Nikitas et al. (2016). Referring to Figure 2.6(a), for both loose samples, there is a clear increase in G under cyclic loading; while for both dense samples, G decreases obviously, which was not observed in the experiments. Reason for this could be due to the irregular shapes of particles, wide particle grading and possible particle abrasion or crushing in experiments, which could lead to continuously densification of samples. However, these phenomena could not be easily replicated in DEM simulations. After 6000 cycles, G of the loose samples and the dense samples at a same σ approach a same constant. As shown in Figure 2.6(b), G increases dramatically with reducing γ_{max} as expected, and in all cases G reduces slightly during cyclic loading. The results correlated quite well with the observations from scaled model tests with different types of offshore wind turbine foundations (Cuéllar et al., 2012; Bhattacharya et al., 2013a,b; Lombardi et al., 2013) and the field measurements (Kühn, 2000).

With the same magnitude of $(\gamma_{max} - \gamma_{min})$ in Series C, the two-way loading resulted in higher G than the one-way loading in the first few hundred cycles (Figure 2.6(c)). The reason is that the true strain level for the one-way loading is larger than those for the two-way loading (i.e. 1.04% > 0.75% > 0.52%), and as expected G decreases with increasing strain level. However, following large number of loading cycles, soil lost memory of initial strain levels and the key parameter dominating the long term stiffness is the magnitude of $(\gamma_{max} - \gamma_{min})$, which are the same for the three simulations, so are the final G values. Similar phenomena also observed in the

DEM simulations of pile-soil interaction under cyclic loading by Cui and Bhattacharya (2016). With the decreasing values of η_c in Series D and E, G decreases obviously as the magnitude of $(\gamma_{max} - \gamma_{min})$ increases (Figures 2.6(d) and 2.6(e)), which confirms the findings from Series C that the key parameter dominating the long term magnitude of stiffness is the magnitude of $(\gamma_{max} - \gamma_{min})$. It can also be observed that G decreases in dense samples and increases in loose samples in general. The exemptions are the dense samples with $\eta_c > 0$, where the magnitude of $(\gamma_{max} - \gamma_{min})$ may be not large enough to produce the same effect.

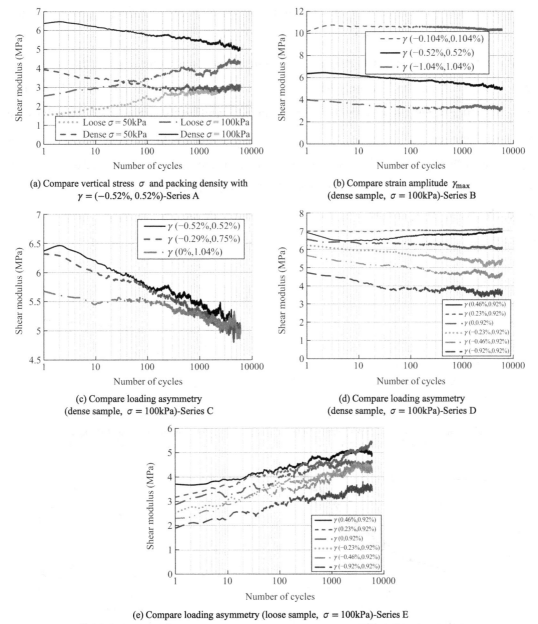

Figure 2.6 Evolution of shear modulus during cyclic loading in DEM simulations (Cui et al., 2019)

The evolutions of void ratio, e, of all simulations under cyclic loading are illustrated in Figure 2.7. The increasing G of loose samples could be explained by the densifications of samples (reduction in e) as observed in Figure 2.7(a). The G values for the dense samples reduce obviously, but they only dilated slightly. There should be other causes for the remarkable decrease of G, which will be investigated further in the latter section. Moreover, e of the two samples at $\sigma = 50\text{kPa}$ coincide, which consists with the coincidence of their G values. However, for the two samples at $\sigma = 100\text{kPa}$, e of the loose sample is still higher than that of the dense sample, which is also consistent with the comparison of their G values. As shown in Figure 2.7(b), with higher $\gamma_{\max}$, dense sample dilates slightly more in the initial stage. However, for $\gamma_{\max} = 1.04\%$, sample dilates first and then contracts slightly. As seen from the monotonic simple shear tests in the latter section (Figure 2.13), this is mainly because the specimen approaches failure from 2% of shear strain thus the sample is disturbed and re-arranged significantly during cyclic loadings. As observed from Series C in Figure 2.7(c), the three simulations with the same value of $(\gamma_{\max} - \gamma_{\min})$ but different values of η_c shows very subtle differences in void ratio. However, with increasing value of $(\gamma_{\max} - \gamma_{\min})$ in Series D/E (Figure 2.7(d)/(e)), void ratio reduces/increases more obviously. This confirms again that the magnitude of $(\gamma_{\max} - \gamma_{\min})$ rather than η_c dominates the responses.

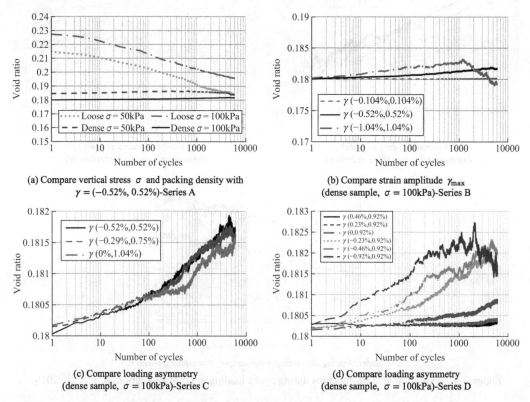

(a) Compare vertical stress σ and packing density with $\gamma = (-0.52\%, 0.52\%)$-Series A

(b) Compare strain amplitude $\gamma_{\max}$ (dense sample, $\sigma = 100\text{kPa}$)-Series B

(c) Compare loading asymmetry (dense sample, $\sigma = 100\text{kPa}$)-Series C

(d) Compare loading asymmetry (dense sample, $\sigma = 100\text{kPa}$)-Series D

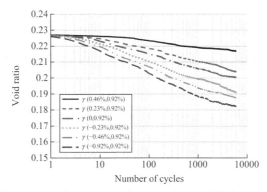

(e) Compare loading asymmetry (loose sample, $\sigma = 100\text{kPa}$)-Series E

Figure 2.7 Evolution of void ratio during cyclic loading in DEM simulations
(Cui et al., 2019)

DEM simulation-micro-scale mechanism

Coordination number

The observed macro-scale stress and strain responses should be underlain by the micro-scale (particle-scale) mechanism. In the following section, micro-scale parameters, including the coordination number, contact force network, fabric anisotropy, principal stress rotation, etc., examined by Cui et al. (2019) to bridge the micro-macro gaps were described.

The coordination number (N_c) is the average number of contacts surrounding each particle. It has a strong relation with the stress level within the sample (Cui et al., 2007). The evolutions of N_c under cyclic loading for the four samples under various test conditions are shown in Figure 2.8. It is clearly shown in Figure 2.8(a) that the initial low N_c corresponds to initial low shear stress (thus low G in Figure 2.6(a)). The increase in G for the two loose samples is related to the increase in N_c and the decrease in G for the dense samples agrees with the reduction in N_c. When values of G coincide, the coordination numbers also coincide. As observed in Figure 2.8(b), with higher $\gamma_{\max}$, N_c reduces more and quicker with loading cycles, matching the reduction of G. As shown in Figure 2.8(c), with the same ($\gamma_{\max} - \gamma_{\min}$), the initial value of N_c was lowest with $\eta_c = 0$ (one-way loading) due to higher $\gamma_{\max}$; however, after 100 cycles, N_c decreases to similar value for all three η_c values, which agrees with the trend for the G in Figure 2.6(c). As seen from Figure 2.8(d), N_c does not change significantly for $\eta_c > 0$ but decrease more significantly with decreasing η_c for $\eta_c \leqslant 0$, which agrees with the observations of G in Figure 2.6(d) and e in Figure 2.7(d). N_c for Series E all increases remarkably with cyclic loading but the differences between various η_c are insignificant.

In summary, considering all observations of G, e and N_c, we can see that lower e (denser packing) results in higher N_c, which in turn leads to higher stress level, thus higher G.

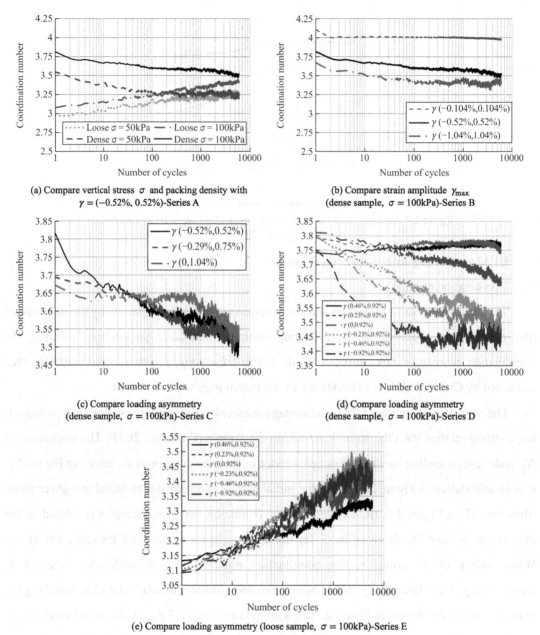

Figure 2.8   Evolution of coordination number (Cui et al., 2019)

<u>Fabric</u>

The spatial distribution of the contact force directions can be quantified by fabric. There is much evidence of the impact of fabric anisotropy on the characteristics of granular materials (Cui and O'Sullivan, 2006; Li et al., 2014). Therefore, it is worth to analyze the evolution of soil fabric in the current cyclic loading conditions. Various definitions of fabric can be found in the literatures (Cambou, 1998). The one adopted in the current study is the Fourier approximation (Rothenburg and Bathurst, 1989), which quantifies the distribution of contact direction per radian as:

$$E(\theta) = \frac{1}{2\pi}[1 + a\cos 2(\theta - \theta_a)] \tag{2.2}$$

where a is a parameter defining the magnitude of fabric anisotropy and θ_a is the direction of the principal fabric. For an isotropic sample, $a = 0$ and $E(\theta) = 1/2\pi$, which is a circle with a uniform distribution of $1/2\pi$ per radian.

The histogram of spatial distribution of directions of particle contact normal for the dense sample at $\sigma = 100\text{kPa}$ is illustrated in Figure 2.9. The Fourier approximation function is indicated by the red ellipse with the long axis indicating the major principal direction of fabric (θ_a). The major principal fabric direction of the initial sample is 92°. When sheared to the maximum strain (0.52% or 0.3°), the major principal fabric direction rotates to the diagonal direction ($\theta_a \approx 130.0°$); when it is sheared to minimum strain (−0.52% or −0.3°), the major principal fabric direction rotates to the perpendicular diagonal direction ($\theta_a \approx 40.0°$). Note that the sample boundaries are only rotated up to 0.6°; however, the major principal fabric direction rotates about 90°.

The evolutions of θ_a at $\gamma_{\max}$ and $\gamma_{\min}$ for Series A and D are plotted in Figure 2.10; the other Series show similar trends to Series A. Note that the initial θ_a at $\gamma = 0$ for the four samples is 83° (loose, $\sigma = 50\text{kPa}$), 165° (loose, $\sigma = 100\text{kPa}$), 80° (dense, $\sigma = 50\text{kPa}$) and 92° (dense, $\sigma = 100\text{kPa}$). It can be observed from Figure 2.10(a), that the rotation of θ_a occurs slowly in loose samples. For the loose sample with $\sigma = 50\text{kPa}$, θ_a at $\gamma_{\max}$ starts from about 110° and increases slowly to 130° at large cycle number; while for the loose sample with $\sigma = 100\text{kPa}$, θ_a at $\gamma_{\max}$ starts from about 140° and decreases slowly to 130°. However, for both dense samples, θ_a at $\gamma_{\max}$ reaches 130° in the first loading cycle and remains at 130°. Rotation of θ_a at $\gamma_{\min}$ shows similar trend: rotation of θ_a occurs slowly in loose samples until it reaches 40°; θ_a in dense samples arrives at 40° in the first cycle. In the other simulation Series, rotation of θ_a does not show obvious difference with various $\gamma_{\min}$ or η_c except for the Simulation D-6 with $\eta_c = 0.5$, where θ_a at $\gamma_{\min}$ does not reduce to 40° as other simulations but stays at 130° for $\gamma_{\min}$ for about 10 cycles, then reduces slowly to 65° following 6000 loading cycles.

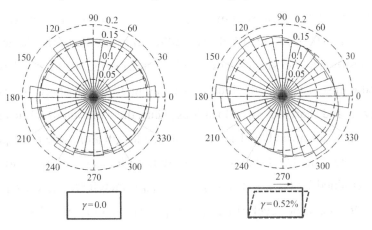

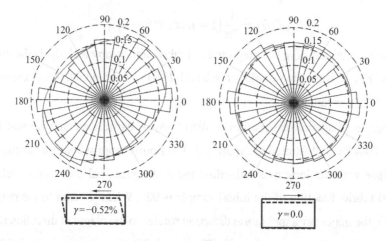

Figure 2.9 Spatial distribution of contact normals (dense sample with $\gamma_{max} = 0.52\%$ and $\sigma = 100$kPa) (Cui et al., 2019)

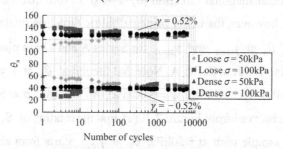

(a) Compare vertical stress σ and packing density with $\gamma = (-0.52\%, 0.52\%)$-Series A

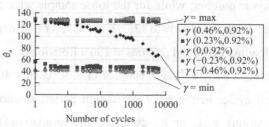

(b) Compare loading asymmetry (dense sample, $\sigma = 100$kPa)-Series D

Figure 2.10 Evolution of major fabric direction at maximum strain and minimum strain (Cui et al., 2019)

The magnitude of fabric anisotropy is quantified by a in Equation (2.2). It has been demonstrated previously (Cui and O'Sullivan, 2006; Li et al., 2014) that larger magnitude of anisotropy can result in higher shear stress. As illustrated in Figure 2.9, a relates to the aspect ratio of the red ellipse. At γ_{max}, the long axis is along 130° and short axis is along 40°; at γ_{min}, the long axis and short axis swop locations. Thus, the variation of fabric anisotropy from γ_{max} to γ_{min} can be quantified by the change of axis length along one principal direction, or $a_{max} + a_{min}$. On the other hand, fabric anisotropy is an indicator of shear stress magnitude, while G is determined by $(\tau_{max} - \tau_{min})/(\gamma_{max} - \gamma_{min})$; therefore, $a_{max} + a_{min}$ could be an indicator of G for the same level of $(\gamma_{max} - \gamma_{min})$. The evolutions of $a_{max} + a_{min}$ for the five simulation Series are shown in Figure 2.11. It is clear in Figure 2.11(a) that $a_{max} + a_{min}$ for the dense samples drops significantly, which agrees with the fact that G for these samples also drops obviously. And $a_{max} + a_{min}$ for the loose samples increases clearly, which matches the increase in G (Figure 2.6(a)). With increasing γ_{max}, degree of anisotropy increases dramatically (Figure 2.11(b)), which reflects the increasing shear stress level. However, as $(\gamma_{max} - \gamma_{min})$ also increases significantly, G reduces. In Series C, $(\gamma_{max} - \gamma_{min})$ remains the same, thus $a_{max} + a_{min}$ for $\eta_c = 0$ is lower than the other two cases in the first 100 cycles and then approaches similar values with other two cases (Figure 2.11(c)), which agrees with the trend for G for this Series (Figure 2.6(c)). In Series D and E, $a_{max} + a_{min}$ increases with decreasing η_c, which results in increasing $(\tau_{max} - \tau_{min})$ (Figure 2.5); however, as $(\gamma_{max} - \gamma_{min})$ also increases with decreasing η_c, the resultant G is actually decreasing. Thus, change of G is a combined result of variation of degree of anisotropy and strain amplitude, and in Series D&E, strain amplitude has a greater impact than the degree of anisotropy.

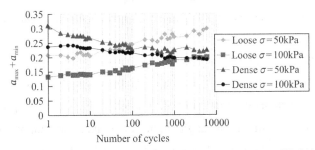

(a) Compare vertical stress σ and packing density with $\gamma = (-0.52\%, 0.52\%)$-Series A

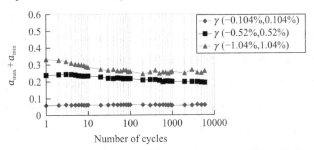

(b) Compare strain amplitude γ_{max} (dense sample, $\sigma = 100$kPa)-Series B

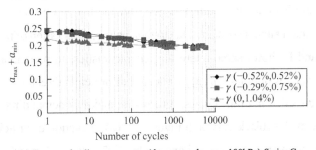

(c) Compare loading asymmetry (dense sample, $\sigma = 100$kPa)-Series C

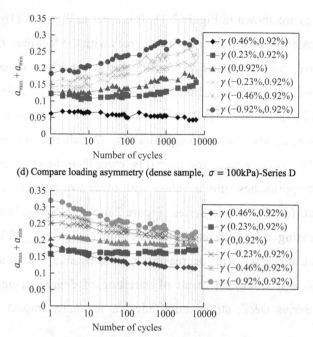

(d) Compare loading asymmetry (dense sample, $\sigma = 100\text{kPa}$)-Series D

(e) Compare loading asymmetry (loose sample, $\sigma = 100\text{kPa}$)-Series E

Figure 2.11   Difference of magnitude of fabric anisotropy between maximum strain and minimum strain (Cui et al., 2019)

<u>Rotation of principal directions</u>

The average stresses within the specimen during cyclic loading were monitored and the principal stresses and principal directions were determined. It is found that the major principal directions and the major fabric directions incline similar angles to the horizontal. The rotations of the major principal directions in the first three cycles were illustrated in Figure 2.12. Note that the initial major principal direction of the loose sample with $\sigma = 100\text{kPa}$ has an approximate horizontal orientation, while the remaining three samples have approximate vertical orientations. The rotation of the major principal direction of the loose sample with $\sigma = 50\text{kPa}$ is slower than the other three samples in the first three cycles (Figure 2.12(a)). The rotation of the major principal direction for various γ_{max} yields same values at maximum and minimum strain but forms larger loops for larger γ_{max} (Figure 2.12(b)). Apart from the loose sample with $\sigma = 50\text{kPa}$, for all the remaining samples with symmetric two-way loading as shown in Figure 2.12(a)-(c), majority of the principal direction rotation occur at about half the γ_{max} (γ_{min}) in the loading (unloading) stages. Outside the strain range of ($\gamma_{min}/2$, $\gamma_{max}/2$), principal directions remain almost constant. Due to this feature, the rotation of principal direction with $\eta_c > 0$ in Series D and E shows some interesting characteristics. In particular, in Simulation D-6, where $\eta_c = 0.5$, as γ_{min} is only half of γ_{max}, rotation of principal direction has not start yet before the loading changes direction in initial cycles. As a result, major principal direction remains at about 130° in the first three cycles (black curves in Figure 2.12(d)). Rotation of principal direction for this sample gradually started after ten cycles and approaches 80° at the end of 6000 loading cycles. This delayed rotation of principal stress direction agrees with the delayed rotation of principal direction of fabric (Figure 2.10(b)). In Simulation E-6 (loose sample with $\eta_c = 0.5$), as sample is relative loose and there are more spaces for particle movements and rearrangements, even the $\gamma_{\min}$ is only half of $\gamma_{\max}$, rotation of principal direction starts from the first cycle to a smaller extent (Figure 2.12(e)). But the shapes of the direction-strain loop are still different from majority simulations: the delay of rotation when loading direction changes at $\gamma_{\min}$ is not observed in Simulation D-6 and E-6. This is because the principal direction has yet stabilized at $\gamma_{\min}$ as other simulations with lower $\gamma_{\min}$.

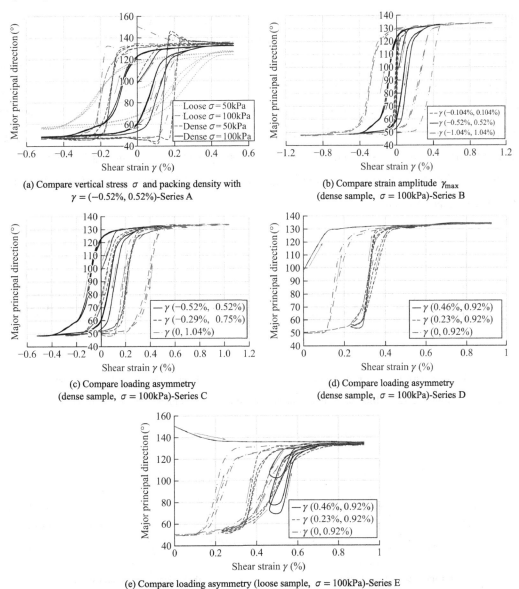

Figure 2.12 Rotation of major principal direction in the first three loading cycles (Cui et al., 2019)

Effect of cyclic loading on p-y curve and critical state analysis

In the design of monopile for the offshore wind turbine, soils surrounding monopile may be replaced by a series of independent springs with their reaction forces applied to monopile specified by *p-y* curves. The *p-y* curve could be attained from the conversion of a τ-γ curve (Zhang and Andersen, 2017). Therefore, it is worth to check the variations of τ-γ curve (thus *p-y* curve) following the cyclic loading.

Figure 2.13 shows the normalized τ-γ curves for the original dense and loose samples with $\sigma = 50$kPa and 100kPa as well as the evolutions of *e* during monotonic simple shear tests until shear strain $\gamma = 52\%$. Dense samples show higher initial stiffness and higher peak stress ratios as expected. Specimen with similar initial *e* shows slightly higher peak stress ratio and initial stiffness under lower σ. The *e* responses of dense samples showed clearly dilation and finally approached critical *e*, while loose samples also showed slightly dilation. Comparing Figure 2.13(b) with Figure 2.7(a), it is interesting to observe that loose specimen densifies during cyclic loading, with *e* approaches constants similar to the values for dense samples; however, these values are not critical *e*. Once these four samples were subjected to 6000 cycles of symmetric loading (e.g., ±0.52%), they were also subjected to monotonic simple shear loading until $\gamma = 52\%$. The stress-strain responses of these four samples given in Figure 2.14 show no significant difference due to the similar values of *e*.

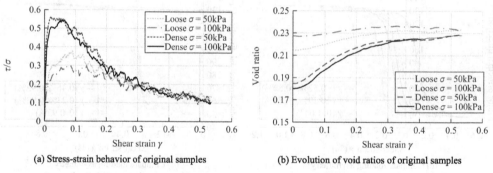

(a) Stress-strain behavior of original samples (b) Evolution of void ratios of original samples

Figure 2.13   Monotonic simple shear tests of original samples (Cui et al., 2019)

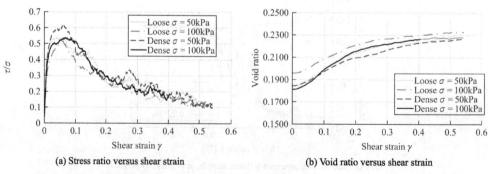

(a) Stress ratio versus shear strain (b) Void ratio versus shear strain

Figure 2.14   Monotonic simple shear tests of cyclically sheared samples (Cui et al., 2019)

To have a better understanding of the relationship of the stress ratios and *e* at the peak and critical states, the stress paths for the eight simulations as well as the critical state line (CSL) were plotted in Figure 2.15, where the mean stress $s' = (\sigma_1' + \sigma_2')/2$, the deviator stress $q = \sigma_1' - \sigma_2'$ and σ_1', σ_2' are the principal stresses. The CSL is obtained by least-square fitting of the e-log s' values or the q-s' values of the eight simulations at the critical state. It can be observed that the e for one relative loose sample ($\sigma = 100$kPa) is just below the critical state e, while the other loose sample ($\sigma = 50$kPa) is below further (Figure 2.15(a)). The values of e for the two dense samples are well below the critical state e. In the simple shear tests, their mean stresses first increase to the peak state with small increases of e, then decrease significantly with further increases of e until reaching the critical state line. It can be seen that the loose sample with $\sigma = 100$kPa has similar response to loose sand; the loose sample with $\sigma = 50$kPa has similar response to medium sand; and the two dense samples behave similar to very dense sand. As shown in Figure 2.15(b), these four samples reach different peak failure envelops due to various initial e.

When the two loose samples subject to cyclic loadings, their e reduce significantly with mean stresses remain unchanged, while the two dense samples remain similar e and mean stresses (Figure 2.15(c)). In the following monotonic simple shear tests, all four samples have similar responses: reaching a same peak failure envelop due to similar e and then moving onto the critical state line (Figure 2.15(d)). Therefore, cyclic loading unifies the distances between the sample state points and the critical state line, thus unifies their responses post cyclic loading.

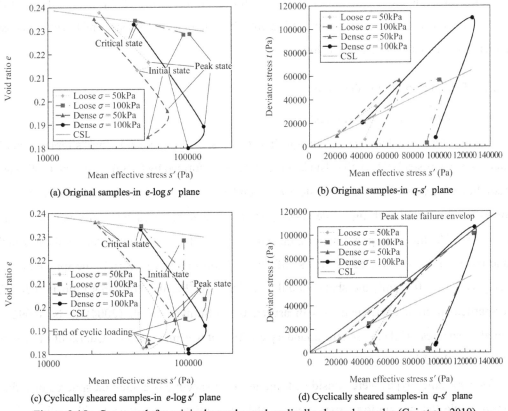

Figure 2.15 Stress path for original samples and cyclically sheared samples (Cui et al., 2019)

2.1.2 Responses of granular soils in cyclic triaxial test conditions

The response of soil under cyclic loading is of great interest to geotechnical engineers. Reflecting the importance of understanding cyclic soil response, much research has been undertaken in this area, and constitutive models for analysis of field scale boundary value problems have been proposed. The complexities of cyclic soil response arise largely due to the particulate nature of soil. Due to the limitations of conventional laboratory testing, much of the experimental research has considered the overall material response by making external measurements on representative samples. While this macro scale approach has helped greatly in developing our understanding of soil response, there is merit in understanding the fundamental particle-scale interactions that underlie the observed macro-scale complexity. The applicability of DEM to develop insight into cyclic soil response has been demonstrated, and DEM was use by O'Sullivan et al. (2008) to develop an understanding of the micro-mechanics of cyclic soil response.

O'Sullivan et al. (2008) coupled a series of relatively simple strain-controlled cyclic triaxial tests with discrete element method (DEM) simulations to quantitatively explore the ability of a DEM model to capture the cyclic response of a granular material. Having validated the DEM model, a parametric study was carried out to examine the influence of the amplitude of cyclic loading on the specimen response, considering both the macro-scale response and the particle-scale interactions. Details are described below.

2.1.2.1 Test and simulation descriptions

Experimental cyclic triaxial test

Many researchers have gained insight into soil response by considering "simple" granular materials, such as steel/glass balls. The laboratory tests performed by O'Sullivan et al. (2008) used assemblies of dry Grade 25 chrome steel balls under vacuum confinement of 80kPa. As these spheres are fabricated with tight tolerances (the sphere diameter and sphericity is controlled to within 7.5×10^{-4}mm during fabrication), the particle geometry can be accurately replicated in the numerical model. As measured by the manufacturer, Thomson Precision Ball, the sphere material density is 7.8×10^3kg/m^3, the shear modulus is 7.9×10^{10}Pa, the Poisson's ratio is 0.28. The inter-particle friction coefficient was measured by O'Sullivan et al. (2004) to be 0.096 for equivalent spheres, while the sphere-boundary coefficient was measured by Cui (2006) in a series of tilt tests to be 0.228.

Two specimen types were considered, the uniform specimens contained spheres of radii of 2.5mm, while the non-uniform specimens contained a mixture of spheres with radii of 2mm,

2.5mm and 3mm in a 1 : 1 : 1 mix. The specimens were 101mm in diameter and 203mm high. The samples were prepared by sealing the latex membrane against the inside of a cylindrical mold using a vacuum. The spheres were then placed using a funnel with a long shaft, the height of the shaft was increased 5 times during the specimen preparation process. The uniform specimens had a void ratio of 0.612, while the non-uniform specimens had a void ratio of 0.603. A representative physical test specimen is illustrated in Figure 2.16(a).

The tests performed by O'Sullivan et al. (2008) were strain controlled, with the axial strain (in compression) being increased to 1% and then reduced to 0%. Due to equipment restrictions, the number of strain cycles in each test was limited to 15 cycles, and these cycles were executed at a rate of 0.0333mm/s by raising and lowering the bottom boundary. The amplitude of the axial strain was 1% in all cases (giving a frequency of cyclic loading of 0.5 cycles per minute). The vertical force applied to the specimen was measured on the stationary top boundary.

DEM Simulations

The DEM simulations performed by O'Sullivan et al. (2008) used a mixed boundary test environment as proposed by Cui et al. (2007). Three different types of boundary condition are used in the "mixed boundary" simulations (refer to Figure 2.16(b)); rigid, planar surfaces are used to simulate the top and bottom platens, a "stress-controlled membrane" is used to simulate the latex membrane that enclosed the specimen laterally in the laboratory, and a set of two, orthogonal vertical periodic boundaries are used so that only one segment of the axi-symmetric specimen need be considered in the simulation. This approach therefore uses the axi-symmetrical property of the triaxial cell to reduce the computational costs of DEM simulations. The only significant disadvantage of this approach is its inability to capture a single shear band passing through the sample as observed, post peak, in triaxial tests on dense or cemented sands. This limitation is not relevant in the context of the described study, the material is relatively loose, and the stress levels considered are lower than the peak stresses mobilized in monotonic tests on this material. To model the flexible latex membrane enclosing the specimen, the confining pressure was applied using a "stress-controlled membrane". The numerical "membrane" is formed by identifying the "membrane spheres", along the outer surface of the sample. A Voronoi diagram of the sphere centroids is used to calculate the force that should be applied to each membrane spheres to achieve the required confining pressure measured in the laboratory, as illustrated in Figure 2.16(c). The force to be applied to each "membrane sphere" is calculated by multiplying the confining pressure and the area of the Voronoi polygon surrounding the sphere centroid. In this test environment, the vertical (deviatoric) load is applied by cycling the rigid top boundaries.

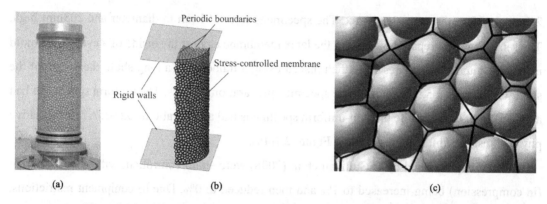

Figure 2.16 Illustration of specimen configuration in laboratory and simulations
(a) Representative laboratory test specimen; (b) Representative "virtual" specimen for DEM analyses (boundary conditions indicated); (c) Subplot of the Voronoi diagram used to simulate membrane in DEM analyses
(O'Sullivan et al., 2008)

The "virtual" one-quarter cylindrical specimen generated for the DEM simulations of the uniform specimens contained 3852 spheres with radii of 2.47mm, with a radius of 50mm and a height of 200mm, and a void ratio of $e = 0.615$. Using a similar approach to the uniform case, a non-uniform specimen containing the same mixture of spheres as in the non-uniform physical tests was developed. This non-uniform specimen contains 3464 non-uniform spheres and had a void ratio of 0.604 (physical test $e = 0.603$).

Considering the input parameters for the DEM simulations, contact was modelled using the elastic Hertz-Mindlin contact model (implementation detailed by Lin and Ng (1997)) with the no-slip tangential contact proposed by Mindlin (1949) being used to calculate shear contact forces. The input parameters for this model are the shear modulus (7.9×10^{10}Pa), and the Poisson's ratio (0.28). The average inter-sphere friction coefficient of 0.096 measured by O'Sullivan et al. (2004) for equivalent spheres was directly input, having established that the results of simulations on randomly packed spheres are insensitive to the small distribution of friction values obtained in these earlier tests. No additional (numerical) damping was applied to the system during the test simulations. As the central difference time-integration system adopted is non-dissipative, energy dissipation in the system occurs via frictional sliding and loss of contact between particles. A scaled density value was used (7.8×10^{12}kg/m^3). Density scaling was adopted to increase the critical time-step and reduce the computational cost of the DEM simulations (see also Thornton (2000), O'Sullivan et al. (2004)). The loading rate in DEM simulations was 100 times smaller the experimental rate to achieve quasi-static condition.

2.1.2.2 Comparison of the laboratory cyclic tests and the DEM simulations

A comparison of the macro-scale response observed in a laboratory test and the DEM simulation for the uniform specimen without area correction is illustrated in Figure 2.17(a), while the non-uniform specimen is considered in Figure 2.17(b), with the deviator stresses normalized by the confining pressures. During cyclic loading hysteresis loops were observed, as is typical for cyclic loading of granular materials, indicating dissipation in energy. Considering both specimens in the first loading cycle the response was somewhat bi-linear with a relatively stiff response (with decreasing stiffness) being observed up until $\varepsilon_a = 0.25\%$ and a significantly less stiff response being observed between $\varepsilon_a = 0.25\%$ and $\varepsilon_a = 1.00\%$. Such a marked change in stiffness was not observed during the subsequent loading cycles as shape of the specimen response during loading evolved with each cycle. For the second loading cycle, the initial stress ratio was about -0.33 for both specimens, but the stress ratio at $\varepsilon_a = 1.00\%$ in this cycle was similar to the stress ratio at $\varepsilon_a = 1.00\%$ in the first cycle, i.e. the overall average stiffness of each specimen in the second cycle was higher than in the first cycle, (the differences in stiffness are related to changes in the specimen fabric below).

During the unloading stages, the stress ratio decreased quickly at the outset of unloading to reach a value of 0 at ε_a values between 0.8% and 0.9%. For all three load cycles considered, the decrease in stiffness during unloading was somewhat bi-linear, with a relatively stiff response observed until $\varepsilon_a = 0.9\%$ and a less stiff response being observed between $\varepsilon_a = 0.9\%$ and $\varepsilon_a = 0\%$. The stress ratio at the end of unloading ($\varepsilon_a = 0\%$) appeared to decrease slightly with increased cyclic loading, with this decrease being less marked for the non-uniform specimen. A good quantitative agreement was attained between the physical test and the simulations with the stress ratio at $\varepsilon_a = 1\%$ being very similar.

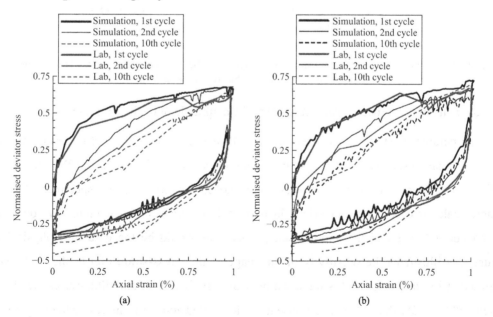

Figure 2.17 Comparison of macro-scale response observed in the laboratory tests and the DEM simulations (without area correction)
(a) Uniform specimen; (b) Non-uniform specimen (O'Sullivan et al., 2008)

Comparisons of the secant stiffnesses for the laboratory tests and the simulations for both the uniform specimen and the non-uniform specimen are shown in Figure 2.18. In both cases the secant stiffness (E_{sec}) calculated was normalized by the confining pressure and E_{sec} was calculated using the stress and strain conditions at the start of the current loading cycle as the origin. The simulation captured the decrease in stiffness with increasing strain relatively accurately for the loading cycles considered here. Considering the loading cycles the numerical simulations indicated a decrease in the rate at which stiffness decreases with increasing axial strain with increasing number of cycles and this was not observed in the physical tests. The shape of stiffness-strain plot in unloading in the first cycle differs significantly from subsequent cycles and this trend is observed in the laboratory tests and the DEM simulations.

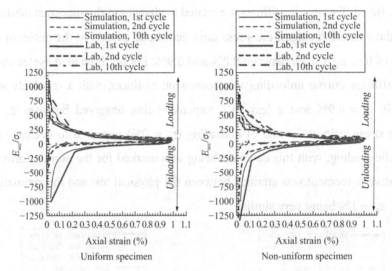

Figure 2.18 Comparison of the secant stiffnesses observed in the laboratory tests and the DEM simulations (cycles 1,2 and 10) (O'Sullivan et al., 2008)

2.1.2.3 Sensitivity of macro-scale response to the amplitude of cyclic straining

The parametric study considered uniform DEM specimens subjected to strain-controlled cyclic triaxial tests with cyclic strain amplitudes of $\varepsilon_a^{max} = 1\%$, $\varepsilon_a^{max} = 0.5\%$, and $\varepsilon_a^{max} = 0.1\%$. Note that all the analyses were carried out on the same specimen. The term "macro-scale" is used here to clearly indicate that these stresses are calculated by considering the boundary forces and the applied cell pressure, as would be the case in a physical test. Figure 2.19 illustrates the sensitivity of the shape of the stress-strain plot to the cyclic strain amplitude (for clarity, for each simulation only the 1st, 2nd, 10th and 50th cycles are shown). When $\varepsilon_a^{max} = 0.5\%$ the bi-linear response in cycle 1 noted above for the case where $\varepsilon_a^{max} = 1\%$ is again evident, while at the lower cyclic strain amplitude of $\varepsilon_a^{max} = 0.1\%$, the decrease in stiffness is smoother and the shape of the response in subsequent cycles is similar to the response observed in the first cycle. The peak deviatoric stress decreased as the maximum strain amplitude decreased, as would be expected.

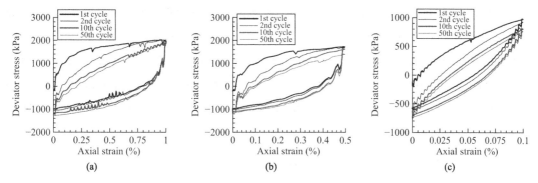

Figure 2.19 Sensitivity of the macro-scale specimen response of uniform sphere specimens in DEM simulations to the cyclic strain amplitude (considering cycles 1,2,10 and 50)
(a) Max. strain-1%; (b) Max. strain-0.5%; (c) Max. strain-0.1% (O'Sullivan et al., 2008)

In the DEM simulations, the variation of a number of parameters during cycling was monitored by considering the situation at the start of each cycle $\varepsilon_a = 0\%$), mid-way through each cycle ($\varepsilon_a = \varepsilon_a^{max}$), at the mid-point of the loading phase and the mid-point of the unloading phase. The resulting data is plotted as a function of the cycle number, n, (starting from $n = 1$) using the following convention; for a given cycle, n, the $\varepsilon_a = 0\%$ data are plotted at point $n - 1$, the $\varepsilon_a = 0.5\varepsilon_a^{max}$ (loading) data are plotted at point $n - 0.75$, the $\varepsilon_a = \varepsilon_a^{max}$ data for unloading are plotted at point $n - 0.5$, and the $\varepsilon_a = 0.5\varepsilon_a^{max}$ (unloading) data are plotted at point $n - 0.25$. As illustrated in Figure 2.20 the deviator stresses at axial strain values of $\varepsilon_a = \varepsilon_a^{max}$, $\varepsilon_a = 0.5\varepsilon_a^{max}$, and $\varepsilon_a = 0$ all tended to decrease as the cyclic loading continued. In the simulations with $\varepsilon_a^{max} = 1\%$ and $\varepsilon_a^{max} = 0.5\%$, there appears to be a consistent gradual decrease in the deviator stress at $\varepsilon_a = \varepsilon_a^{max}$. In contrast for these simulations when $\varepsilon_a = 0$, and for $\varepsilon_a = 0.5\varepsilon_a^{max}$ (both loading and unloading) there was a notable decrease in the deviator stress over the first 3 cycles, followed by a more gradual decrease as cyclic loading progressed. For the simulation with $\varepsilon_a^{max} = 0.1\%$ there was a significant decrease in the deviator stress at $\varepsilon_a = \varepsilon_a^{max}$ over the initial 10 cycles, in comparison with a more gradual decrease observed in the subsequent 40 cycles. For this simulation, as for the other two simulations, the decrease in deviator stress at $\varepsilon_a = 0.5\varepsilon_a^{max}$ was more noticeable in comparison with the stress decrease at $\varepsilon_a = \varepsilon_a^{max}$. Referring to Figure 2.17, this decrease in the deviatoric stress values during loading, results in a reduction in the area of the hysteresis loop and reduction in the amount of energy dissipated in each cycle as loading continues.

The variation in normalized secant stiffness (E_{sec}/σ_3) during the 1st, 2nd and 10th cycles for each simulation considered is illustrated in Figure 2.21. Considering the data presented in Figure 2.21, there is no noticeable differences in the variation in stiffness as a function of strain during subsequent cycles (i.e. the response observed in the 10th cycle is indistinguishable from the response observed in the 50th cycle). The fluctuations observed in the response at low levels of straining (Figure 2.21, $\varepsilon_a^{max} = 0.1\%$) are a consequence of the elastic nature of the contact springs used in the rheological model used to represent contact in the DEM model. These changes in stiffness correspond to the small fluctuations observed in the stress-strain plots (Figure 2.17). These small fluctuations will have had little influence on the overall response, we chose not to remove them by filtering. Such a response would not be observed in a physical granular material and highlights the need to implement a dissipative contact model to study granular material response at very small strains.

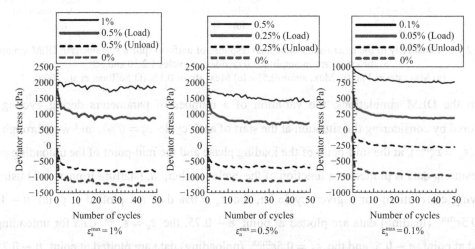

Figure 2.20 Sensitivity of macro scale response to strain amplitude in DEM simulations of cyclic loading, as a function of the number of load cycles (O'Sullivan et al., 2008)

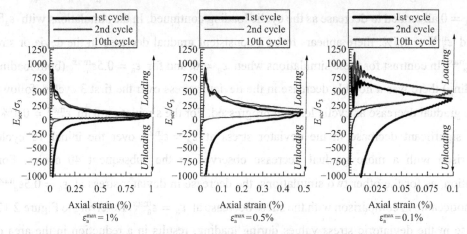

Figure 2.21 Evolution of the secant stiffness with strain for the three simulations considered (cycles 1,2 and 10) (O'Sullivan et al., 2008)

2.1.2.4 Evolution of micro-scale parameters during cyclic loading

Evolution of contact force network

Diagrams of the contact force network and plots indicating the magnitudes and orientations of contact forces at the middle and the end of the 50th cycle are illustrated in Figures 2.22 for the simulations with $\varepsilon_a^{max} = 1\%$ and $\varepsilon_a^{max} = 0.1\%$. For these diagrams, only the strong contact forces, i.e. those contact forces exceeding the average contact force plus one standard deviation are considered. For all three simulations these strong contact forces make up 15% of the total number of contacts (and this proportion does not vary significantly during the simulations). In Figure 2.22 lines are drawn between the centers of contacting spheres, and the line thickness is proportional to the magnitude of the force, the entire specimen volume is considered here. For the case where $\varepsilon_a^{max} = 1\%$, most of the large contact forces were oriented in the vertical direction at the maximum axial strain value (Figure 2.22(a)), while the largest contact forces were oriented in the horizontal direction at axial strain of 0% (Figure 2.22(c)). At $\varepsilon_a^{max} = 0.1\%$, while the contact forces are also clearly aligning themselves with the orientation of the maximum principal stress (Figure 2.22(b)), for this smaller strain amplitude the anisotropy in the contact force network is less marked at $\varepsilon_a = 0\%$ (Figure 2.22(d)).

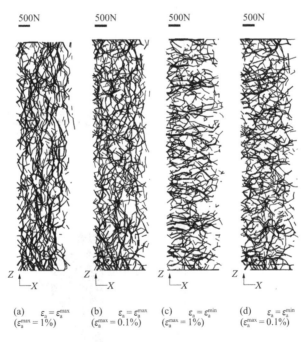

(a) $\varepsilon_a = \varepsilon_a^{max}$ (b) $\varepsilon_a = \varepsilon_a^{max}$ (c) $\varepsilon_a = \varepsilon_a^{min}$ (d) $\varepsilon_a = \varepsilon_a^{min}$
($\varepsilon_a^{max} = 1\%$)   ($\varepsilon_a^{max} = 0.1\%$)   ($\varepsilon_a^{max} = 1\%$)   ($\varepsilon_a^{max} = 0.1\%$)

Figure 2.22 Contact force network at maximum and minimum strain levels in 50th cycle for uniform specimen
(considering only forces > average force + 1 std. dev.)
(O'Sullivan et al., 2008)

Plots of the contact force network such as those provided in Figure 2.22 can give only a qualitative assessment of the arrangement of the network of contacts transmitting stress through the material given the complexity of the network and its three-dimensional geometry. A quantitative assessment of the contact forces may be made by reference to the polar histogram plots provided as Figures 2.23 to 2.24. All non-zero contact forces were considered in the development of these plots. As the specimen is axi-symmetric we need consider only one quadrant of the system to plot the histograms. As would be expected for this axi-symmetric system, the distribution of contact forces orientations in the horizontal plane is approximately uniform, therefore only the vertical projections are considered here. Each $10°$ bin in the histogram has been shaded and the degree of shading indicates the average contact force magnitude in that bin, normalized by the overall average contact force for the strain level considered. Therefore these plots give an indication of both the orientation of the contact forces in the system, as well as the relative magnitudes of the forces transmitted in each direction.

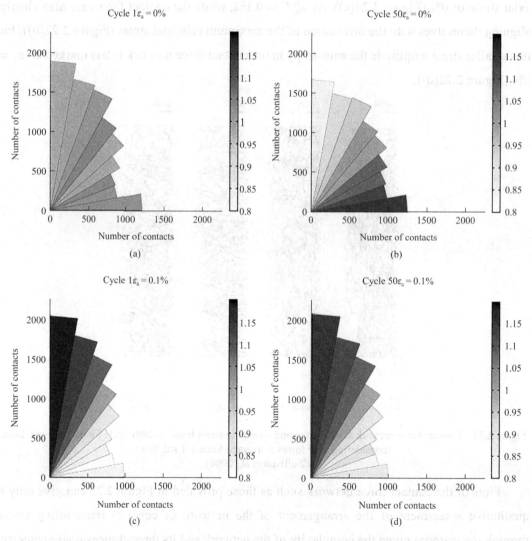

Figure 2.23 Polar histograms of contact force orientation in cycles 1 and 50 for $\varepsilon_a^{max} = 0.1\%$, shading illustrates normalized contact force magnitudes (O'Sullivan et al., 2008)

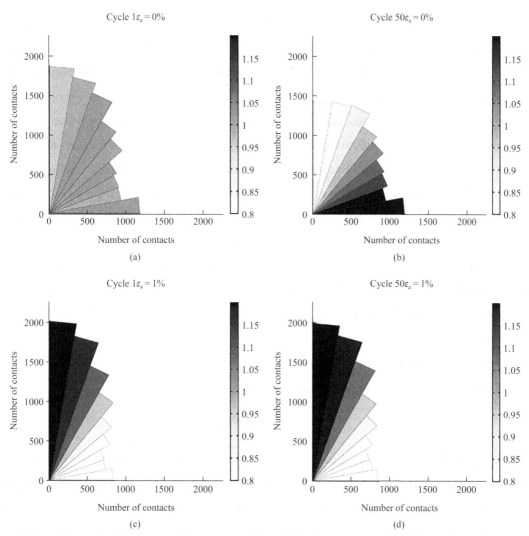

Figure 2.24 Polar histograms of contact force orientation in cycles 1 and 50 for $\varepsilon_a^{max} = 0.1\%$, shading illustrates normalized contact force magnitudes (O'Sullivan et al., 2008)

Figures 2.23 and 2.24 consider all the contacts in the specimen for the simulations with $\varepsilon_a^{max} = 0.1\%$ and $\varepsilon_a^{max} = 1\%$ respectively. In both figures the distribution of the contact forces at the beginning and end of the first and last load cycles simulated is considered. Comparing firstly the distribution of contact force orientations, initially, as a consequence of the specimen generation approach there are more contact normals orientated vertically (i.e. with an inclination to the horizontal exceeding 45°) than horizontally. This anisotropy in the contact orientation is more pronounced when $\varepsilon_a = \varepsilon_a^{max}$ in both simulations, and there is a slight difference in the distribution with the simulation where $\varepsilon_a^{max} = 0.1\%$ having more horizontally orientated contacts in comparison with the simulation where $\varepsilon_a^{max} = 1\%$. A more notable difference between the two simulations is the distribution in the magnitude of contact forces as a function of contact orientation. In the simulation with $\varepsilon_a^{max} = 0.1\%$, the forces orientated in the vertical direction (i.e. > 80° to the horizontal) tend to be about 1.15 times the average force, however where $\varepsilon_a^{max} = 1\%$, the forces tend to be closer to 1.2 times the average force and the normalized horizontal force magnitudes are smaller. Looking at the difference between the contact force distributions at the beginning and end of the last load cycle it is clear that as the orientation of the major principal stress rotates, and the deviator stress moves from a positive to a negative value (refer to Figure 2.20), there is a significant difference in the contact force orientations as well as the relative magnitudes of the forces transmitted in the horizontal and vertical directions. Considering Figure 2.23(b) and Figure 2.24(b) we can see that when the orientation of the major principal stress is horizontal (negative deviator stress), while the distribution of contact orientations is almost uniform there are still, on average, more forces orientated vertically. However the horizontally orientated contacts transmit more force than the vertically orientated contacts. When the major principal stress is orientated vertically the distribution of forces is considerably more anisotropic, with about 70% of the contacts having an orientation exceeding 50° to the horizontal in both cases.

Fabric tensor analysis

For spherical particles, the fabric tensor is given by

$$\Phi_{ij} = \frac{1}{2N_c} \sum_{k=1}^{N_c} n_i^{(k)} n_j^{(k)} \tag{2.3}$$

where N_c is the number of contacts, n_i is the component of the unit branch vector in the i direction, and the branch vector is the vector joining the centroids of the two contacting particles. The principal values, Φ_1, Φ_2 and Φ_3, and the principal directions of the fabric tensor can be calculated by considering the eigenvalues and eigenvectors of the fabric tensor. The deviator fabric $\Phi_1 - \Phi_3$) quantifies the anisotropy of the microstructure (see also Thornton (2000) and Cui and O'Sullivan (2006)). For the three simulations considered the fabric tensor was calculated firstly by considering all the contacts in the specimen and then by considering only the strong contacts (i.e. contacts where the contact force exceeded the average force + 1 standard deviation).

The evolution of the overall anisotropy (i.e. $\Phi_1 - \Phi_3$ considering all the contacts) for the duration of each of the three simulations considered is illustrated in Figure 2.25. A comparison of Figures 2.25 and 2.20 reveals a clear link between the variation in deviator stress and the evolution of fabric anisotropy. Considering firstly the deviator fabric at 0% strain, initially $\Phi_1 - \Phi_3$ is almost zero, indicating an isotropic fabric, corresponding with the initial isotropic stress condition. Comparing the simulations with $\varepsilon_a^{max} = 1\%$ and $\varepsilon_a^{max} = 0.5\%$, there is a more noticeable decrease in deviator stress at $\varepsilon_a = \varepsilon_a^{max}$ for the $\varepsilon_a^{max} = 0.5\%$ simulation (Figure 2.20) and this corresponds with a more noticeable decrease in deviator fabric (Figure 2.25).

Considering the response at $\varepsilon_a = 0$, for all three simulations a significant decrease in deviator stress over the initial cycles of loading (i.e. magnitude increased) is observed to correspond with a significant increase in anisotropy. Note also that as cyclic loading continued the deviator fabric at $\varepsilon_a = 0$ exceeded the deviator fabric at $\varepsilon_a = \varepsilon_a^{max}$ for all three simulations, even though the deviator stress at $\varepsilon_a = \varepsilon_a^{max}$ exceeded the deviator stress at $\varepsilon_a = 0$ (This is explored further below). The variation in fabric with increased cyclic loading appears more marked than the variation in the deviator stress. This suggests that the fabric does not depend only upon the deviator stress but also on the previous loading of the specimen. The deviator fabric is clearly continuing to evolve as cycling progresses and has not reached a stable state after 50 load cycles, even when $\varepsilon_a^{max} = 0.1\%$.

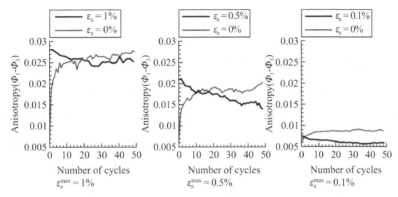

Figure 2.25 Evolution of specimen fabric anisotropy during 50 load cycles
(O'Sullivan et al., 2008)

Recognizing that there is significant heterogeneity in the contact force network and considering the strong force chains that are typically observed in two dimensional DEM analyses, the fabric tensor was re-evaluated, considering only the contacts transmitting the largest contact forces (i.e. forces exceeding the average contact force plus one standard deviation). The deviator fabric in this case, i.e. the "strong force" deviator fabric, is illustrated in Figure 2.26. Note that following the definition of mechanical coordination number proposed by Thornton (2000), the fabric tensor was also calculated considering only contacts where each sphere meeting at that contact point had two or more contacts, however the deviator fabric did not noticeably change for any of the simulations, in comparison with the values presented in Figure 2.25. However, the strong force deviator fabric, illustrated in Figure 2.26, clearly correlates more strongly with the stress data in Figure 2.20, in comparison with the overall deviator fabric illustrated in Figure 2.25. The deviator fabric at $\varepsilon_a = 0$ never exceeds the strong force deviator fabric at $\varepsilon_a = \varepsilon_a^{max}$, and the variation in strong force deviator fabric with increased cyclic loading more closely resembles the variation in the deviator stress with increased cyclic loading.

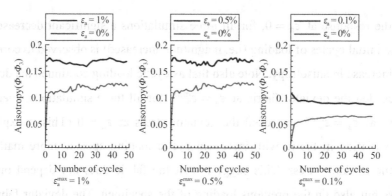

Figure 2.26 Evolution of specimen strong force fabric anisotropy during 50 load cycles (considering only forces > average force + 1 std. dev.) (O'Sullivan et al., 2008)

Coordination number evolution

The coordination number was calculated as

$$N = 2N_c/N_p \tag{2.4}$$

where N_c is the number of contacts and N_p is the number of particles. Figure 2.27 illustrates the variation in N as a function of axial strain for selected loading cycles, while Figure 2.28 illustrates the variation of N as a function of the number of cycles at selected strain levels. Referring to Figure 2.27, for each simulation the N value decreased significantly in the first cycle, this is consistent with the significant decrease in coordination number at small strains observed in the monotonic triaxial simulations of Cui et al. (2007). The magnitude of the reduction in the coordination number decreased with decreasing values of ε_a^{max}. The variations in coordination number in subsequent cycles were noticeably smaller. The coordination number data clearly then indicates that the biggest change in the specimen fabric took place in the first cycle, correlating with the significant changes in the macro scale response in the first cycle (Figure 2.20).

It is interesting to observe that the maximum coordination number occurred when ε_a was close to $0.5\varepsilon_a^{max}$ during the loading stage, and not at the maximum ε_a value. Referring to Figure 2.28, after the initial load cycles (approximately 5) the coordination numbers tended to increase as cyclic loading continued. Values of N can be related to the observed macro-scale response, the significant decrease in N at $\varepsilon_a = 0$ in the first cycles of loading corresponded with a significant decrease in deviator stress. As noted above the area of the hysteresis loops decreases slightly as cyclic loading continues, energy will be dissipated in friction and also as particles loose contact. Comparing the variation in coordination number with axial strain for cycles 2 and 50 for all three simulations, there is no apparent reduction in the number of contacts that are broken and reformed in a given load cycle as the number of cycles of loading increases.

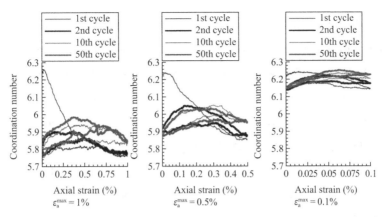

Figure 2.27 Variation in coordination number with strain for various amplitudes of cyclic loading (cycles 1,2,10 and 50) (O'Sullivan et al., 2008)

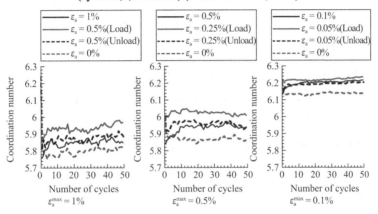

Figure 2.28 Evolution of coordination number with number of load cycles (O'Sullivan et al., 2008)

2.1.3 Summary

Based on the experimental tests performed by Nikitas et al. (2016) and DEM simulations performed by Cui et al. (2019), the following conclusions were arrived:

- Shear modulus for loose soil increases rapidly in the initial loading cycles as a result of soil densification, and then the rate of increase diminishes when void ratio approaches a constant; shear modulus of the very dense samples in DEM decreases but dilation is not the main reason.
- Under the same vertical stress and strain amplitude, shear moduli and void ratios of dense sample and loose sample approach the same values at larger number of cycles.
- Shear modulus increases with increasing vertical stress and relative density but decreases with increasing strain amplitude as expected.
- Higher shear stress level and shear modulus are correlated to higher coordination number and higher magnitude of fabric anisotropy.
- Accumulated rotation of principal direction of fabric and major principal direction

occurs slowly in loose sample but reaches the final value in the first cycle in dense sample. In each symmetric loading cycle, majority of the principal direction rotation occurs between half of γ_{min} and half of γ_{max} in spite of the magnitude of γ_{max}. There are delays in the reverse of principal direction rotation immediately after the boundary rotation is reversed apart from the case with $\eta_c = 0.5$ where the vibration intensity is low, and sample has not stabilized at γ_{min}.

- Loading asymmetry only affects soil behaviors in the first hundreds of cycles. In long term, the magnitude of ($\gamma_{max} - \gamma_{min}$) rather than loading asymmetry dominates the soil responses.

- Changes in the stress-strain behavior (thus p-y curve) of soil due to cyclic loading are significantly dependent on the initial relative density and the distance between the state point and the critical state line.

Based on the coupled physical cyclic triaxial tests and DEM simulations performed by O'Sullivan et al. (2008), the DEM model was proved to be able to capture the responses observed in the physical tests. The following conclusions were arrived:

- There is a decrease in the deviator stress as calculated using macro-scale parameters as cyclic loading progressed for all simulations, and this decrease is most noticeable over the initial load cycles.

- Following an initial increase in the calculated secant stiffness values, the secant stiffness at the maximum strain values remained approximately constant, while there was an apparent decrease in the secant stiffness at $\varepsilon_a = 0.5\varepsilon_a^{max}$.

- The distribution of contact force orientations and magnitudes reflects the macro-scale applied stress state. There is a clear redistribution in the magnitude of contact force during cyclic loading, with the contacts carrying the largest forces tending to be orientated in the direction of the major principal stress. The variation in the number of contacts orientated in the major principal direction is less marked.

- Comparing the principal stress differences and the deviator fabric, the stress measurements correlated better with the strong fabric tensor (i.e. the fabric tensor calculated using only the contacts transmitting the largest contact forces). At $\varepsilon_a = \varepsilon_a^{max}$ most of the stress is transmitted through the system by only 15% of the total number of contacts. The strong fabric tensor varied less with increased cyclic loading, in comparison with the overall fabric tensor.

- The coordination number at $\varepsilon_a = \varepsilon_a^{max}$ drops significantly in the 1st cycle then tends to increase slightly as cyclic loading continues, the variation in coordination number at

$\varepsilon_a = 0$ is less obvious. While a stiffer material response tends to coincide with a higher coordination number, the deviator fabric relates more strongly with the stress strain response observed at the macro scale, than does the coordination number.

- The parametric study has illustrated that even under relatively small amplitude cyclic loading ($\varepsilon_a^{max} = 0.1\%$), the specimen fabric continues to evolve as cyclic loading continues, a steady state is not achieved after 50 cycles of loading.

2.2 Cyclic behavior of China Laizhou Bay submarine mucky clay at an offshore wind turbine site

The Laizhou Bay submarine soils are mainly mucky soft clay. The previous in-door laboratory investigations on submarine soils concentrated on the dynamic characteristics in terms of the static and dynamic strength (Narasimha Rao & Panda, 1998; Aghakouchak et al., 2015), cumulative strain development (Vucetic & Dobry, 1988; Ren, Xu, Teng, et al., 2018), pore pressure development (Lee & Sheu, 2007; Leng et al., 2018) and modulus attenuation (Gu et al., 2017; Yang et al., 2017; Leng et al., 2018). In particular, Zhang et al. (2018) and Wang et al. (2019) studied the influence of medium principal stress and consolidation process on the shear strength of Shanghai and Yantai submarine soils. Wichtmann et al. (2013) investigated the effects of amplitude and loading frequency on the cyclic strength of Norwegian submarine clay by cyclic triaxial and direct shear tests. Leng et al. (2017) studied the influence of soil structure on the development of strain and effective stress path through undrained dynamic cyclic triaxial tests on Shanghai submarine soft clay. Wang et al. (2013) studied the development of cumulative strain and pore pressure of Wenzhou submarine soft clay under cyclic load at different confining pressure levels.

The bearing capacity reduction under cyclic loadings is a critical consideration for the foundation design of offshore wind turbines (Liu et al., 2019; Ashour & Norris, 2000; Dai et al., 2021), where the cyclic induced excess pore pressure is the main reason. Therefore, researchers studied the cumulative development law of pore water pressure of submarine soils (Wang et al., 2017). In particular, Hyde et al. (1994) studied the development of pore pressure of submarine sediment through cyclic triaxial tests. Li et al. (2011), Moses et al. (2003) and Wang et al. (2018) studied the cumulative process of pore pressure under cyclic loads and its effect on post-cyclic shear strength. Ren et al. (2018) studied the pore pressure development law of submarine soft clay, and proposed a hyperbolic function model based on the laboratory results.

Considering the intrinsic feature of large number of cycles for wave loads, there is a lack of laboratory investigations of submarine soils under such special load condition (usually smaller than 2000 cycles for existing studies). In addition, the investigation of cyclic characteristics and failure mechanism of submarine soft clay in Laizhou Bay is very limited. Therefore, this study concentrates on the investigation of the mechanical behavior of a submarine clay at Laizhou Bay, China, a potential construction site for a 700MW wind farm. Cyclic triaxial tests are carried out to reveal the cyclic characteristics of this silty soft clay. The laboratory experiments analyze the cyclic stress-strain and stress path behavior, revealing the cyclic mobility behavior under different Cyclic Stress Ratio (CSR) and buried depth levels. The buried depth refers to the vertical distance of the samples from the sea bed level. Further research investigates the insight of mechanical behavior by concentrating on the development laws of cumulative strain, pore pressure and modulus attenuation behavior, highlighting the difference between the failure and non-failure scenarios. Furthermore, based on the experimental results, critical CSR values are obtained against buried depth, as the failure-triggering threshold load boundary. This provides critical reference for the dynamic stability evaluation and design for monopile foundations under cyclic wave loads for wind power constructions in silty soft clay.

2.2.1 Experiment process

2.2.1.1 Testing apparatus and soil samples

The GDS dynamic triaxial testing apparatus was used for the experiments, as shown in Figure 2.29. The tested soil samples were taken from Changyi City, Shandong Province, China. A 700MW wind farm is under construction in Changyi City. According to the statistics of other researchers (Gao, 2012), the average sea water depth is 18m, and the average wave period is 2.5-3.8s. Seventy-five 4MW wind turbines and a 220kV offshore booster station will be installed in the first phase of the project. Sixty-four 6MW wind turbines and a 220kV offshore booster station will be installed in the second phase of the project. After the completion of the project, the annual power generation capacity will reach 1.75 billion kWh.

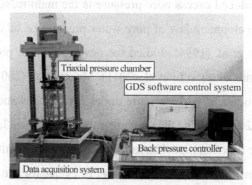

Figure 2.29 GDS dynamic triaxial testing system

Intact samples were obtained according to ASTM D1587-15 (2015), *Standard Practice for Thin-Walled Tube Sampling of Fine-Grained Soils for Geotechnical Purposes*. The buried depth of the samples ranges from 2 to 11m b.g.l. (below ground level). The submarine surficial soils contain 2% organic matter and a small amount of impurities such as shells. The liquid/plastic limits were tested by the combined method of liquid-plastic limit according to ASTM D4318-17e1 (2015), *Standard Test Methods for Liquid Limit, Plastic Limit, and Plasticity Index of Soils*. The specific gravity was tested by water pycnometer according to ASTM D854-14 (2015), *Standard Test Methods for Specific Gravity of Soil Solids by Water Pycnometer*. The moisture content was obtained by the drying method according to ASTM D2216-19 (2015), *Standard Test Methods for Laboratory Determination of Water (Moisture) Content of Soil and Rock by Mass*. The test results showed that the average liquid limit, plastic limit, plastic index, and compressibility of the soil samples were 48.0%, 22.5%, 25.5, and 1.50MPa^{-1} respectively. Furthermore, the natural void ratio of the tested clay was found to be 1.466, and the natural water content was found to be 52.8%. According to *Code for investigation of geotechnical engineering* (GB 50021—2019), the tested soil sample in this study is a submarine mucky soft clay. The grain size distribution curves of the samples at different depths are identical, and the C-3-3 one at 6m b.g.l. is shown in Figure 2.30 as an example. The coefficient of nonuniformity (C_u) and the curvature coefficient (C_c) were 2.4 and 0.8167 respectively, indicating the unfavorable particle gradation. The obtained physical properties are shown in Table 2.3.

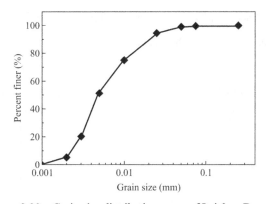

Figure 2.30 Grain size distribution curve of Laizhou Bay clay

The fundamental physical properties of mucky clay　　　　　Table 2.3

w (%)	ρ (g/cm³)	ρ_d (g/cm³)	e_0	Liquid limit w_L (%)	Plastic limit w_P (%)	Plastic index IP	G_s
52.8	1.71	1.12	1.466	48.0	22.5	25.5	2.76

Cylindrical samples were cut by wire saw and soil cutter to a size with 50mm in diameter and 100mm in height, as shown in Figure 2.31(a). The tested soil samples were taken from submarine boreholes, with high water content and high saturation, where the initial B values of the samples are very close to 1. The vacuum saturation method was adopted as the only saturation method in this study, and the saturation of soil samples reached the test requirement after 24 hours vacuum saturation. The samples were submerged in a vacuum saturator after cutting, and were placed on the apparatus immediately after the completion of saturation, which could reduce sample disturbance. Furthermore, the underwater condition in the vacuum saturator is also consistent with the submarine environment on site (see Figure 2.31(b)). Figure 2.31(c) shows the sample installation.

(a) Sample cutting (b) Sample saturation (c) Sample installation

Figure 2.31 Preparations of soil samples (Han et al., 20??)

The soil samples were firstly tested monotonically under CU condition, subjected to three confining pressures, i.e. 100kPa, 200kPa and 300kPa. The tests were completed when axial strain reached to 15%-20%. During the test, the specimen was compressed and deformed, the middle part expanded, and no obvious failure surface appeared. The c' and φ' values were derived from the Mohr-Coulomb graph (Figure 2.32), with $c' = 10.09$ and $\varphi' = 31.2°$.

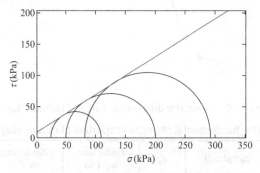

Figure 2.32 The Mohr-Coulomb graph

2.2.1.2 Testing scheme

The soil samples tested in this paper were all undisturbed samples from the site. For the soil samples from different buried depths in this study, a series of compression tests were carried out.

Pressures were prescribed to 10kPa, 30kPa, 50kPa, 100kPa, 200kPa, 300kPa, 400kPa, 600kPa, 800kPa, 600kPa, 400kPa, 200kPa, 100kPa, 50kPa, 100kPa, 200kPa, 300kPa, 400kPa, 600kPa, 800kPa, 1600kPa in sequent. Consolidation was kept for 24 hours at each stage of the applied pressures, and the e-p curves are illustrated in Figure 2.33. The confining pressures in the triaxial tests were determined according to the corresponding effective self-weight and depth of the tested soil samples, ranging 15kPa-70kPa. The consolidation pressure of each layer is listed in Table 2.4 pre-consolidation stress was approximately 80kPa. The pre-consolidation stresses were higher than the confining pressures in the tests, and therefore the soil specimens were over-consolidated clay. For group A, confining pressure was varied to simulate the different buried depths of the soil samples. At a specific depth, cyclic stress ratio (CSR) was varied to investigate the impact of wave amplitude, where five levels (0.3-0.5) were specified (Leng et al., 2018; Moses et al., 2003; Wang et al., 2017; Aghakouchak et al., 2015). In particular, CSR is defined as follows,

$$\text{CSR} = \frac{P}{\sigma_c} \tag{2.5}$$

where P is the amplitude of applied cyclic stress and σ_c is the effective confining pressure of the soil specimen. For group B, the cyclic stress amplitude was kept at constant to concentrate on the impact of buried depth on the dynamic soil behavior. For all the tests, the loading frequency was maintained as 0.33Hz, i.e.3s, according to the average wave period in the Laizhou Bay.

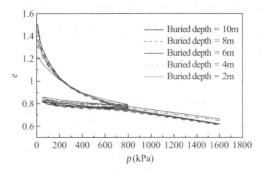

Figure 2.33 The e-p curve

2.2.1.3 Testing procedure

During the test, the advanced-loading module of the GDS testing apparatus was adopted. This module can realize the application and control of the confining pressure, axial pressure and back pressure of the sample. The B value, defined as the ratio between induced pore water pressure and incremental confining pressure, $\frac{\Delta u}{\Delta \sigma}$, was used to quantify the saturation degree of soil samples. 98% was adopted as the lower limit for B values of all the tests to guarantee the saturation status of samples. Otherwise, the back-pressure saturation should be continued until the saturation meets the requirements. Then, soil samples were isotopically consolidated under the defined consolidation pressures for different cases, as listed in Table 2.4. A filter paper was pasted around the edge of the specimen to assist consolidation. Considering the disturbance of samples, every effort was made to reduce and avoid disturbance during all the transport, sampling and testing elements. For example, the soil samples were wrapped in iron sheets and plastic wrap to prevent extrusion and evaporation. In the consolidation process, in order to reduce sample disturbance, back pressure and cell pressure were applied to the internal and external boundaries of the samples respectively. During the tests, the difference between the back pressure value and the cell pressure value was monitored to consolidate the samples, and this can reduce the disturbance from the external pressure. The specific drainage volume of each sample in the consolidation stage was about 7000-10000mm^3, accounting for 0.35%-0.51% of the sample volume. Therefore, the volumetric strains of samples in the consolidation stage were found to be very small, and the soil samples were basically undisturbed in the consolidation stage. During the test, the pore water pressure of the sample was measured at the bottom of the sample. The completion of the consolidation process was recognized when pore water pressure stabilized to the back pressure. Static triaxial tests were carried out under the undrained condition and cyclic tests were conducted for samples under different confining pressures and CSR levels. In particular, in the static tests, strain control mode was employed, and the shearing rate was 0.05mm/min. Stress control mode was adopted in the cyclic tests, and the loading frequency was controlled as 0.33Hz. The criteria for defining the phenomenon of liquefaction are commonly defined as 90% pore water pressure ratio or 5% axial strain. The development of the concerned parameters (such as pore pressure, axial strain, deviator stress, etc.) was recorded during the tests.

Summary of the dynamic triaxial tests   Table 2.4

Test group	Test number	Depth (m)	Moisture content (%)	Confining pressure (kPa)	CSR	Cyclic stress amplitude (kPa)
Group A	C-1-1	2	56.78	15	0.3	4.5
	C-1-2	2	56.78	15	0.4	6.0
	C-1-3	2	56.78	15	0.5	7.5
	C-1-4	2	56.78	15	0.665	10.0
	C-1-5	2	56.78	15	0.7475	11.2
	C-2-1	4	55.69	25	0.3	7.5
	C-2-2	4	55.69	25	0.4	10.0
	C-2-3	4	55.69	25	0.45	11.3
	C-2-4	4	55.69	25	0.475	11.9
	C-2-5	4	55.69	25	0.5	12.5
	C-3-1	6	50.14	40	0.3	12.0
	C-3-2	6	50.14	40	0.35	14.0

Table 2.4 (Continued)

Test group	Test number	Depth (m)	Moisture content (%)	Confining pressure (kPa)	CSR	Cyclic stress amplitude (kPa)
Group A	C-3-3	6	50.14	40	0.375	15.0
	C-3-4	6	50.14	40	0.4	16.0
	C-3-5	6	50.14	40	0.5	20.0
	C-4-1	8	47.96	55	0.3	16.5
	C-4-2	8	47.96	55	0.35	19.5
	C-4-3	8	47.96	55	0.375	20.6
	C-4-4	8	47.96	55	0.4	22.0
	C-4-5	8	47.96	55	0.5	27.5
	C-5-1	10	41.36	70	0.3	21.0
	C-5-2	10	41.36	70	0.325	22.8
	C-5-3	10	41.36	70	0.35	24.5
	C-5-4	10	41.36	70	0.4	28.0
	C-5-5	10	41.36	70	0.5	35.0
Group B	C-6-1	2	56.78	15	0.83	12.5
	C-6-2	4	55.69	25	0.50	12.5
	C-6-3	6	50.14	40	0.31	12.5
	C-6-4	8	47.96	55	0.23	12.5
	C-6-5	10	41.36	70	0.18	12.5

2.2.2 Cyclic behavior

Cyclic triaxial tests were carried out to investigate the cyclic characteristics of the Laizhou Bay silty soft clay, including the stress-strain and stress path behavior. This is to reveal the cyclic mobility behavior under different CSR and buried depths. Further research investigates the insight mechanical behavior by concentrating on the development law of cumulative strain and pore pressure and the modulus degradation behavior, highlighting the difference between the failure and non-failure scenarios. Furthermore, based on the experimental results, critical CSR values were obtained against buried depth, as the failure-triggering threshold load boundary.

2.2.2.1 Stress-strain behavior

At the end of consolidation stage, there was a small axial strain observed (0.3-0.4%) for all samples. The insignificant axial deformation of all samples was uniformly cleared prior to the cyclic tests. The stress-strain behavior of the soil samples at the buried depth of 4m and 8m are shown in Figure 2.34(a)-(c) show the cyclic behavior at buried depth 4m subjected to CSR levels of 0.45, 0.475 and 0.5 respectively. It can be seen that when CSR increases from 0.45 to 0.475, no substantial difference is induced. In particular, both samples show similar stress-strain behavior of narrow loops and slight stiffness degradation behavior as number of cycles increases. Neither of the samples fails after 10000 load cycles. However, when CSR increases to 0.5, the hysteretic behavior becomes significant and the induced deformation sharply increases after approximately 300 cycles, where a sample failure is observed after reaching 600 cycles. The stress-strain behavior shows alternatively strain softening and hardening behavior, due to contraction and dilation of the Laizhou Bay clay under cyclic loads. Increasing number of cycles induces profound stiffness degradation and damping increase due to the loop inclination and area change respectively. These observations indicate the existence of a critical CSR value, which defines the threshold for triggering cyclic failure of the investigated Laizhou Bay clay. Figure 2.34(d)-(f) show the cyclic behavior at buried depth 8m subjected to CSR levels of 0.3, 0.35 and 0.5 respectively. The impact of CSR on the cyclic behavior is similarly observed as the case of 4m. More significant stiffness degradation and damping increase are induced under higher CSR, resulting to more significant soil failure. However, a smaller critical CSR boundary value is observed for the soil at deeper buried depth. This will be discussed in detail in Section 2.2.2.3. Figure 2.35 shows the final state of three specimens (8m b.g.l.) under CSR levels of 0.3, 0.35 and 0.5. In particular, the soil sample failure under CSR 0.5 is observed, which agrees with the observations from the stress strain behavior.

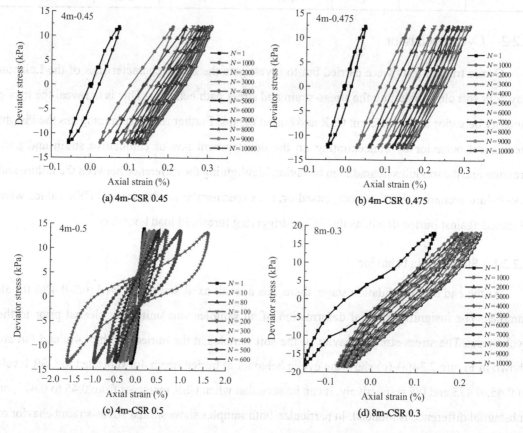

(a) 4m-CSR 0.45 (b) 4m-CSR 0.475

(c) 4m-CSR 0.5 (d) 8m-CSR 0.3

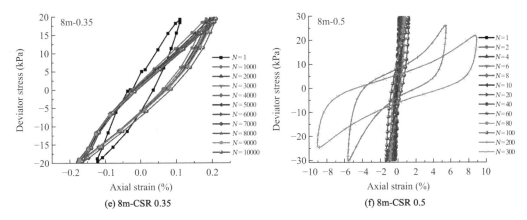

Figure 2.34 Cyclic stress-strain behavior

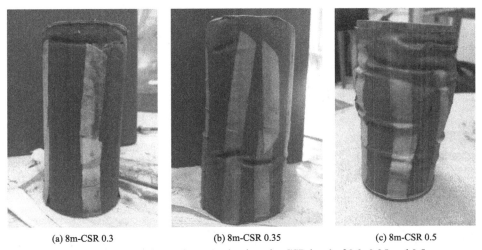

(a) 8m-CSR 0.3   (b) 8m-CSR 0.35   (c) 8m-CSR 0.5

Figure 2.35 Tested samples at 8m b.g.l. under CSR level of 0.3, 0.35 and 0.5

2.2.2.2 Stress path behavior

Figure 2.36 shows the stress path behavior of the soil samples at the buried depth of 4m and 8m. In particular, Figure 2.36(a)-(c) show the cyclic behavior at buried depth 4m subjected to CSR levels of 0.45, 0.475 and 0.5 respectively. It can be seen that all three samples show the effective stress reduction under cyclic loads, while when under CSR level of 0.5, the stress state reaches the failure envelope and oscillates along the compression & extension portions passing instantaneously through the origin. This triggers the initial liquefaction and leads to significant development of cyclic deformation and finally sample failure. However, when subjected to the CSR 0.45 and 0.475, there is no liquefaction observed, where the stress path is far below the initial liquefaction point. This is in agreement with the observations from the stress strain behavior. For the cyclic soil behavior at the buried depth of 8m, when CSR level exceeds 0.35, there is a significant decrease of effective stress, which leads to the occurrence of liquefaction (cyclic mobility) and soil failure.

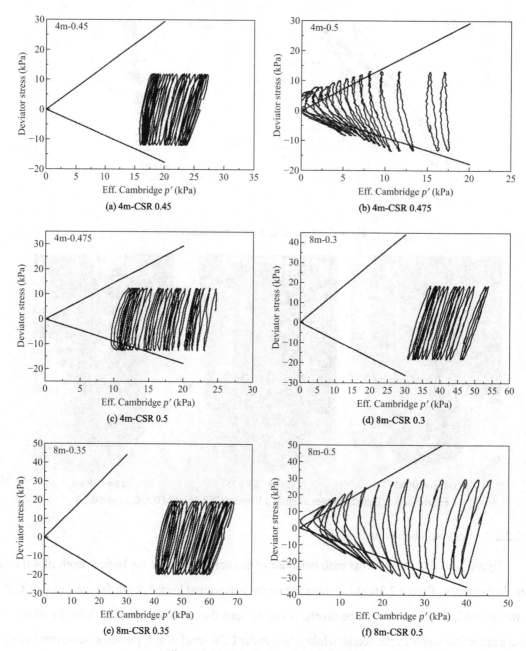

Figure 2.36   Cyclic stress path behavior

2.2.2.3   Development of cumulative strain

<u>Development of cumulative strain affected by different CSR</u>

The cumulative axial strain development against number of cycles is shown in Figure 2.37 under different buried depths and CSR levels. Figure 2.37(a) shows the soil behavior at 4m b.g.l. For low CSR levels (i.e. 0.3, 0.4, 0.45 and 0.475), as number of cycles increases, cumulative strain gradually increases and no destructive deformation is induced. Furthermore, higher CSR levels leads to more significant axial strain development. However, for high CSR level (i.e. 0.5), a sharp increase of axial strain is observed when number of cycles reaches approximately 300, and sample failure is induced at around 600 cycles. Additional tests were carried out to establish the CSR threshold for triggering soil failure, i.e. the critical CSR value under which there is a substantial increase of cyclic deformation. This value is found to be 0.4875 for the case of 4m b.g.l.. For the soil behavior under 6m, 8m and 10m b.g.l., similar behavior is observed. The observations also highlight the different strain development laws between the failure and non-failure scenarios. In particular, when below the critical CSR value, cumulative axial strain increases gently with number of cycles. When above the critical CSR value, after a smooth accumulation, axial strain increases sharply with number of cycles after a turning point.

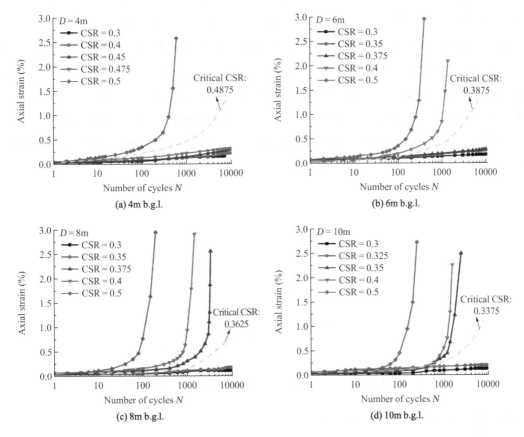

Figure 2.37 Cumulative strain development under different CSR

According to above dynamic triaxial test results, the failure-triggering critical cyclic pressures against depths was obtained, as illustrated in Table 2.5. Four tests of soil specimens under confining pressures of 25kPa, 40kPa, 55kPa and 70kPa were carried out to determine the undrained shear strength, and the failure-triggering critical cyclic pressures was normalized by the undrained shear strength.

Table 2.5 Critical state of soil specimens at various depths

Depth (m)	Confining pressure (kPa)	Critical cyclic pressure (kPa)	Undrained shear strength (kPa)	Normalized critical pressure
4	25	12.1875	57	0.21
6	40	15.5	66	0.23
8	55	19.9375	73	0.27
10	70	23.625	81	0.29

Development of cumulative strain affected by different depths

The development of cumulative strain against number of cycles at different depths is shown in Figure 2.38, i.e. case C-6-1 to C-6-5 in Table 2.4. The buried depths of the five specimens are 2m, 4m, 6m, 8m and 10m respectively, and the unified cyclic stress amplitude is kept as 12.5kPa for all the five cases. The results show that under the same stress amplitude, the shallower soil sample is more prone to cyclic failure, indicating its lower strength. In particular, as shown in Figure 2.38, the cumulative strain of the specimen at 2m b.g.l. increases the fastest, and its failure turning point appears at the smallest number of cycles. The specimen at 4m b.g.l. shows similar behavior, where a breaking turning point and the subsequent strain leap is observed. The cumulative strain of the specimens at 6m, 8m and 10m b.g.l. increases more gently as the number of cycles increases, and no failure is observed.

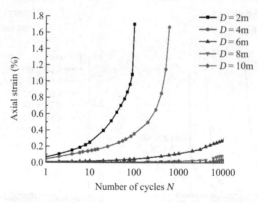

Figure 2.38 Results of cumulative strain under different depths

2.2.2.4 Development of pore pressure

Development of pore pressure affected by different CSR

During the tests, pore pressure was measured at the bottom of samples. The variation of pore pressure ratio, i.e. the ratio between instantaneous cyclic pore pressure and effective confining pressure, against number of cycles is investigated considering the impact of CSR level and buried depth, as shown in Figure 2.39.

Taking the results at 4m b.g.l. as an example, for low CSR levels (i.e. 0.3, 0.4, 0.45 and

0.475), cyclic induced pore water pressure ratio increases more significantly in the first 4000 cycles, and then stabilizes. No failure is observed for these samples. Higher CSR level leads to more profound development of pore water pressure. However, for high CSR level (i.e. 0.5), a significant development is observed for pore water pressure ratio, which reaches 1.0 after approximately 600 cycles. This results in the cyclic liquefaction (cyclic mobility) and finally the soil failure. This is in agreement with the observations from the cumulative strain development in the previous section. The failure-triggering threshold CSR value is observed as 0.4875, identical with the one obtained from the strain behavior. Similar pore water pressure development behavior is observed for other cases, while it is found that the critical CSR value decrease with deeper buried depth.

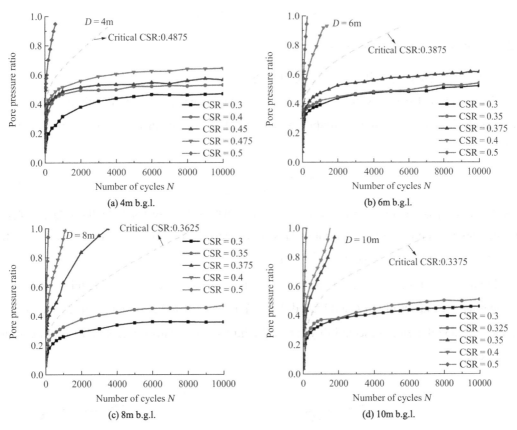

Figure 2.39 Pore pressure ratio development under different CSR

Development of pore pressure ratio affected by different depths

The development of pore pressure at different depths is shown in Figure 2.40. Five specimens were tested under the same cyclic stress amplitude (12.5kPa) but at different buried depths of 2m, 4m, 6m, 8m and 10m (case C-6-1 to C-6-5 in Table 2.4 respectively). The test results show that under the same cyclic stress amplitude, shallower soil experiences faster development of pore pressure, indicating its lower cyclic resistance. The pore pressure growth follows approximately a hyperbolic growth trend.

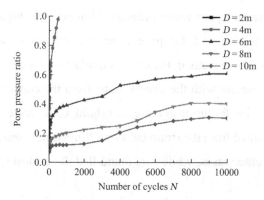

Figure 2.40   Results of pore pressure ratio under different depth

2.2.2.5   Stiffness attenuation behavior

<u>Stiffness attenuation affected by different CSR</u>

The discussed stiffness in the section is the secant shear modulus of each hysteretic loop by considering the increasing number of cycles. The secant shear modulus is defined as follows,

$$E_d = \frac{\sigma_{max} - \sigma_{min}}{\varepsilon_{dmax} - \varepsilon_{dmin}} \tag{2.6}$$

where σ_{max}, σ_{min} are the maximum and minimum shear stresses in a specific loop, ε_{dmax}, ε_{dmin} are the maximum and minimum shear strains in a specific loop. The shear modulus ratio is the secant shear modulus normalized by its initial value.

Figure 2.41 shows the secant modulus attenuation behavior against number of cycles, considering the impact of CSR and buried depth. Larger CSR induces more significant stiffness degradation and therefore susceptibility for cyclic liquefaction. This trend is more significant for deeper soil.

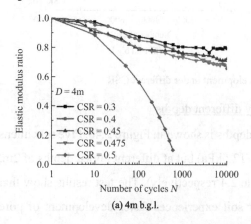

(a) 4m b.g.l.

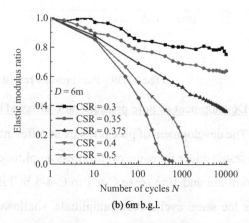

(b) 6m b.g.l.

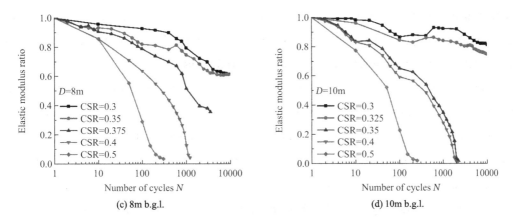

Figure 2.41 Shear modulus attenuation under different CSR

<u>Stiffness attenuation affect by different depths</u>

In Figure 2.42, when subjected to the same cyclic stress amplitude, the stiffness attenuation shows different behavior under different buried depths. In particular, the shallower soil sample show more profound secant modulus attenuation behavior. Two groups of specimens with buried depth of 2m and 4m were destroyed when the tests were carried out to 100 and 600 cycles respectively. The specimens with buried depth of 6m, 8m and 10m were not destroyed in the later stage of the tests due to the lower dynamic stress level, but the modulus of the specimens decreases in varying degrees.

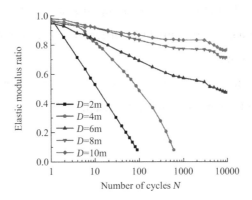

Figure 2.42 Results of shear modulus ratio under different depth

2.2.3 Summary

This study investigates the dynamic characteristics of Laizhou Bay submarine clay under long-term cyclic loads. Cyclic triaxial tests were carried out to reveal the cyclic characteristics of this silty soft clay. The laboratory experiments analyzed the cyclic stress-strain and stress path behavior, revealing the cyclic mobility behavior under different Cyclic Stress Ratio (CSR) and buried depth levels. The insight of mechanical behavior of this material was studied by concentrating on the development laws of cumulative strain, pore pressure and stiffness attenuation behavior, highlighting the difference between the failure and non-failure scenarios. Some of the experimental findings from the study are summarized as follows.

(1) Stress-strain behavior of Laizhou Bay clay under low CSR levels shows narrow loops and slight stiffness degradation as number of cycles increases. The stress-strain behavior under high CSR levels shows alternatively strain softening and hardening behavior, due to soil contraction and dilation under cyclic loads. Increasing number of cycles induces more profound stiffness degradation and pore pressure increase.

(2) Stress state under high CSR levels reaches the failure envelope and oscillates along the compression & extension portions passing instantaneously through the origin. This triggers the initial liquefaction and leads to significant development of cyclic deformation and finally sample failure.

(3) Experimental observations indicate the existence of a critical CSR value, which defines the threshold for triggering cyclic failure of the investigated Laizhou Bay clay. The results highlight the different strain development laws between the failure and non-failure scenarios. When below the critical CSR value, cumulative axial strain increases gently with number of cycles. When above the critical CSR value, after a smooth accumulation, axial strain increases sharply with number of cycles after a turning point.

(4) For the shallower soil samples, the confining pressure p_0 in the cyclic tests is smaller. Therefore, the stress ratio of testing soil samples is different, which would result in more obvious failures in the shallower soil samples.

(5) Larger CSR induces more significant stiffness degradation and therefore susceptibility for cyclic liquefaction. This trend is more significant for deeper soil.

References

Aghakouchak A, Sim W W, Jardine R J. Stress-path laboratory tests to characterise the cyclic behaviour of piles driven in sands. Soils and Foundations, 2015, 55: 917-928.

Andersen K H. Bearing capacity under cyclic loading-offshore, along the coast, and on land. The 21st Bjerrum Lecture presented in Oslo, 23 November 2007. Canadian Geotechnical Journal, 2009, 46(5): 513-535.

API. Recommended practice for planning, designing, and constructing fixed offshore platforms: working stress design, RP2A-WSD, 20th edn. Washington, DC: American Petroleum Institute, 1993.

Ashour M, Norris G. Modeling Lateral Soil-Pile Response Based on Soil-Pile Interaction. Journal of Geotechnical and Geoenvironmental Engineering, 2000, 126: 420-428.

ASTM D1587-15. Standard Practice for Thin-Walled Tube Sampling of Fine-Grained Soils for Geotechnical Purposes, West Conshohocken, PA, USA: ASTM International, 2015.

ASTM D2216-19.Standard Test Methods for Laboratory Determination of Water (Moisture) Content of Soil and Rock by Mass, West Conshohocken, PA, USA: ASTM International, 2015.

ASTM D4318-17e1. Standard Test Methods for Liquid Limit, Plastic Limit, and Plasticity Index of Soils, West Conshohocken, PA, USA: ASTM International, 2015.

ASTM D6528-07. Standard Test Method for Consolidated Undrained Direct Simple Shear Testing of Cohesive Soils, 2007.

ASTM D854-14. Standard Test Methods for Specific Gravity of Soil Solids by Water Pycnometer, West Conshohocken, PA, USA: ASTM International, 2015.

Bhattacharya S, Cox J A, Lombardi D, et al. Dynamics of offshore wind turbines on two types of foundations. Proceedings of the Institution of Civil Engineers: Geotechnical Engineering, 2013a, 166(GE2), 159-169.

Bhattacharya S, Nikitas N, Garnsey J, et al. Observed dynamic soil-structure interaction in scale testing of offshore wind turbine foundations. Soil Dynamics and Earthquake Engineering, 2013b, 54, 47-60.

Cambou B. Micromechanical approach in granular mechanics. In: Cambou B. (ed.) Behaviour of Granular Materials. number 385 in CISM Courses and Lectures. Springer-Verlag, Wien New York, 1998.

Cuéllar P, Georgi S, Baeßler M, et al. On the quasi-static granular convective flow and sand densification around pile foundations under cyclic lateral loading. Granular Matter, 2012, 14(1), 11-25.

Cui L, Bhattacharya S, Nikitas G, A Bhat Micromechanics of granular soil in asymmetric cyclic loadings: an application to offshore wind turbine foundations. Granular Matter, 2019, 21.

Cui L, Bhattacharya S. Soil-monopile interactions for offshore wind turbines. Proceedings of the ICE-Engineering and Computational Mechanics, 2016, 169(4): 171-182.

Cui L, O'Sullivan C, O'Neil S. An analysis of the triaxial apparatus using a mixed boundary three-dimensional discrete element model. Geotechnique, 2007, 57(10): 831-844.

Cui L, O'Sullivan C. Exploring the macro-and micro-scale response characteristics of an idealized granular material in the direct shear apparatus. Geotechnique, 2006, 56(7): 455-468.

Cui L. Developing a virtual test environment for granular materials using discrete element modelling. PhD. Thesis, University College Dublin, Ireland, 2006.

Dai S, Han B, Li N, Wang B, et al. Morphologic analysis of hysteretic behavior of China Laizhou Bay submarine mucky clay and its cyclic failure criteria. Bulletin of Engineering Geology and the Environment, 2022, 81: 52.

Dai S, Han B, Wang B, et al. Influence of soil scour on lateral behavior of large-diameter offshore wind-turbine monopile and corresponding scour monitoring method. Ocean Engineering, 2021, 239: 109809.

DNV. Offshore standard: Design of offshore wind turbine structures, DNV-OS-J101. Hellerup, Denmark: Det Norske Veritas, 2004.

Doherty P, Gavin K. Laterally loaded monopile design for offshore wind farms. Proceedings of the Institution of Civil Engineers-Energy, 2012, 165: 7-17.

Gao B. The numerical simulation and analysis of wave field in Laizhou Sea. Qingdao: Ocean University of China, 2012.

Gu C, Gu Z, Cai Y, et al. Dynamic modulus characteristics of saturated clays under variable confining pressure. Canadian Geotechnical Journal, 2017, 54: 729-735.

Hyde A F L, Yasuhara K, Hirao K. Stability criteria for marine clay under one-way cyclic loading. Journal of Geotechnical Engineering, 1994, 119(11): 1771-1789.

Itasca PFC2D 4.00 Particle Flow Code in Two Dimensions: User's Guide. Itasca Consulting Group, Inc., Minnesota, USA, 2008.

Jalbi S, Arany L, Salem A, et al. A method to predict the cyclic loading profiles (one-way or two-way) for monopile supported offshore wind turbines. Marine Structures, 2019, 63: 65-83.

Kelly R B, Houlsby G T, Byrne B W. Transient Vertical Loading of Model Suction Caissons in a Pressure Chamber. Geotechnique, 2006, 56(10): 665-675.

Kühn M. Dynamics of offshore wind energy converters on monopile foundations-experience from the Lely offshore wind farm. OWEN Workshop "Structure and Foundations Design of Offshore Wind Turbines" March 1, 2000, Rutherford Appleton Lab.

Leblanc C, Houlsby G T, Byrne B W. Response of stiff piles in sand to long-term cyclic lateral loading. Géotechnique, 2010, 60: 79-90.

Lee C J, Sheu S F. The stiffness degradation and damping ratio evolution of Taipei Silty Clay under cyclic straining. Soil Dynamics and Earthquake Engineering, 2007, 27: 730-740.

Leng J, Liao C, Ye G, et al. Laboratory study for soil structure effect on marine clay response subjected to cyclic loads. Ocean Engineering, 2018, 147: 45-50.

Leng J, Ye G, Ye B, et al. Laboratory test and empirical model for shear modulus degradation of soft marine clays. Ocean Engineering, 2017, 146: 101-114.

Li L L, Dan H B, Wang L Z. Undrained behavior of natural marine clay under cyclic loading. Ocean Engineering, 2011, 38: 1792-1805.

Li X, Yu H S. Fabric, force and strength anisotropies in granular materials: a micromechanical insight. Acta Mechanica, 2014, 225(8): 2345-2362.

Lin X, Ng T T. A three-dimensional discrete element model using arrays of ellipsoids. Geotechnique, 1997, 47(2): 319-329.

Liu B, Jeng D S, Ye G L, et al. Laboratory study for pore pressures in sandy deposit under wave loading. Ocean Engineering, 2015, 106: 207-219.

Liu J W, Cui L, Zhu N, et al. Investigation of cyclic pile-sand interface weakening mechanism based on large-scale CNS cyclic direct shear tests. Ocean Engineering, 2019, 194: 106650.

Lombardi D, Bhattacharya S, Muir Wood D. Dynamic soil-structure interaction of monopile supported wind turbines in cohesive soil. Soil Dynamics and Earthquake Engineering, 2013, 49: 165-180.

Lombardi D, Bhattacharya S, Scarpa F, et al. Dynamic response of a geotechnical rigid model container with absorbing boundaries. Soil Dynamics and Earthquake Engineering, 2015, 69: 46-56.

Mindlin R D. Compliance of elastic bodies in contact. Transactions of the ASME, Series E, Journal of Applied Mechanics, 1949, 20(327): 221-227.

Moses G G, Rao S N, Rao P N. Undrained strength behaviour of a cemented marine clay under monotonic and cyclic loading. Ocean Engineering, 2003, 30: 1765-1789.

Narasimha Rao S, Panda A P. Non-linear analysis of undrained cyclic strength of soft marine clay. Ocean Engineering, 1998, 26: 241-253.

Nikitas G, Arany L, Aingaran S, et al. Predicting long term performance of offshore wind turbines using cyclic simple shear apparatus. Soil Dynamics and Earthquake Engineering, 2017: 73.

O'Sullivan C, Bray J, Riemer M. An examination of the response of regularly packed specimens of spherical particles using physical tests and discrete element simulations. Journal of Engineering Mechanics, ASCE, 2004, 130(10): 1140-1150.

O'Sullivan C, Cui L, O'Neil S. Discrete element analysis of the response of granular materials during cyclic loading. Soils and Foundations, 2008, 48(4): 511-530.

Ren X W, Xu Q, Teng J, et al. A novel model for the cumulative plastic strain of soft marine clay under long-term low cyclic loads. Ocean Engineering, 2018, 149: 194-204.

Ren X W, Xu Q, Xu C B, et al. Undrained pore pressure behavior of soft marine clay under long-term low cyclic loads. Ocean Engineering, 2018, 150: 60-68.

Rothenburg L, Bathurst R J. Analytical study of induced anisotropy in idealized granular materials.

Geotechnique, 1989, 39: 601-614.

Thornton C. Numerical simulations of deviatoric shear deformation of granular media. Geotechnique, 2000, 50(1): 43-53.

Vucetic M, Dobry R. Degradation of Marine Clays under Cyclic Loading. Journal of Geotechnical Engineering, 1988, 114: 133-149.

Wang J H, Ling X, Li Q, et al. Accumulated permanent strain and critical dynamic stress of frozen silty clay under cyclic loading. Cold Regions Science and Technology, 2018, 153: 130-143.

Wang J, Guo L, Cai Y, et al. Strain and pore pressure development on soft marine clay in triaxial tests with a large number of cycles. Ocean Engineering, 2013, 74: 125-132.

Wang Y K, Gao Y, Guo L, et al. Cyclic response of natural soft marine clay under principal stress rotation as induced by wave loads. Ocean Engineering, 2017, 129: 191-202.

Wang Y Z, Lei J, Wang Y, et al. Post-cyclic shear behavior of reconstituted marine silty clay with different degrees of reconsolidation. Soil Dynamics and Earthquake Engineering, 2019, 116: 530-540.

Wichtmann T, Andersen K H, Sjursen M A, et al. Cyclic tests on high-quality undisturbed block samples of soft marine Norwegian clay. Canadian Geotechnical Journal, 2013, 50: 400-412.

Yang Z, Yuan J, Liu J, et al. Shear Modulus Degradation Curves of Gravelly and Clayey Soils Based on KiK-Net In Situ Seismic Observations. Journal of Geotechnical and Geoenvironmental Engineering, 2017, 143: 06017008.

Zhang S, Ye G, Liao C, et al. Elasto-plastic model of structured marine clay under general loading conditions. Applied Ocean Research, 2018, 76: 211-220.

Zhang Y, Andersen K H. Scaling of lateral pile p-y response in clay from laboratory stress-strain curves. Marine Structures, 2017, 53: 124-135.

Zhao X. The change of wave dynamic environment for Laizhou Bay in recent 30 Years. Tianjin: Tianjin University of Science and Technology, 2013.

Zhu B, Byrne B W, Houlsby G T. Long-Term Lateral Cyclic Response of Suction Caisson Foundations in Sand. ASCE Journal of Geotechnical and Geoenvironmental Engineering, 2013, 139(1): 73-83.

Chapter 3 Driving characteristics of open-ended pile for offshore wind power

The pile foundation, as one of the most common forms of deep foundations, are used widely in the fields of engineering construction, such as industrial and civil construction, road, bridge and port engineering, ocean engineering, etc. At present, the monopile foundation is the most widely used foundation type for offshore wind farms all over the world, which is usually an open-ended steel tubular pipe pile with a diameter of 3-8m. In the process of pile driving, part of soil is squeezed into the pipe pile and formed "soil plug". The soil plug effect is the main factor that distinguishes the open-ended pipe pile from the closed-ended pipe pile or solid pile, and the main reason for the different bearing characteristics between them. Research on the formation of soil plug and the load transfer mechanism in the process of pile driving is the key to accurately predict the driving behavior of open-ended pipe piles. However, the installation of pile shoe and the change of pile diameter will inevitably change the formation of soil plug, and then change the driving characteristics and bearing performance of open-ended pipe piles. Therefore, this chapter study the mechanical mechanism of the driving process of open-ended pile under different pile shoe forms and pipe pile diameters based on large-scale model test combined with discrete element modeling, in order to provide a reliable theoretical basis for accurately predicting the construction behavior of open-ended pile.

3.1 Large scale model test on driving characteristics of open-ended pile

3.1.1 Test equipment and material selection

3.1.1.1 Large scale model test

The laboratory model test consists of four parts: model box system, loading system, loading control system and data acquisition system, as shown in Figure 3.1. The inner dimension of the model tank is 3000mm × 3000mm × 2000mm (length × width × height). The loading system consists of hydraulic cylinder, high pressure oil pump, pressure control box and PLC (Programmable Logic Controller) control system. Qingdao sea sand (dry sand) was selected for the experiments, and the sieving method was used to determine the particle size distribution of the soil sample, as shown in Figure 3.2. Other parameters of the soil sample were provided in Table 3.1.

Dry sand was poured into the soil tank in layers of 100mm each to ensure the homogeneity, and finally the sand bed reached a height of 1800m. The relative density of sand sample is 73%, which is in the dense state. The miniature fiber Bragg grating strain sensor was used to monitor the pile strain during the whole process of pile driving; the YWD-100 displacement sensor was used to dynamically monitor the vertical displacement of the ground around the pile; the MPS cable displacement sensor was used to record the pile settlement and soil plug height in real time, and the pressure sensor was installed on the pile top o record the pile driving resistance.

Figure 3.1　Large scale model test　　　　Figure 3.2　Particle size distribution curve

Parameters of sand samples　　　　　　　　　　　　　　　Table 3.1

Relative density G_s	Maximum void ratio e_{max}	Minimum void ratio e_{min}	Average grain size d_{50} (mm)	Particle size range (mm)	Internal frictional angle φ (°)
2.65	0.52	0.30	0.72	0-15	42.8

3.1.1.2　Model pile and layout

Pipe pile with double-layered wall is an effective way to capture the inner and outer frictional resistance of open-ended pile simultaneously. Paik & Lee (1933) firstly used double-walled model pipe pile to carry out load test. Subsequently, some other researchers used double-walled pipe pile to study the driving behavior of pile (Lehane & Gavin, 2001; Choi & O'Neill, 1997; Gavin & Lehane, 2003), and have achieved valuable results. However, there are few reports on the influence of pile shoe form on the driving characteristics of open-ended piles. In this section, double-walled pipe piles are used to carry out the research.

The double-walled pile model consists of two concentric pipes of 6063 aluminum alloy material. The outer diameter, wall thickness and the pile length are 140mm, 13mm and 1065mm respectively. Yegian et al. (1973) proved through finite element analysis and Rao et al. (1996) proved through model test that the boundary effect can be ignored when the boundary of the model box is outside the $6D$-$8D$ range of the pile body. Therefore, the boundary effect can be ignored in the pile driving process for the model box and model pile selected in this study. The tops of the inner and outer pile are connected by bolts, and the sensitized microfiber grating sensors are installed in the groves on the outside surface of the outer pipe and on the outside surface of the inter pipe, respectively. The schematic diagram of double-walled pipe pile is shown in Figure 3.3, the layout of model pile sensors is shown in Figure 3.4, and the physical picture is shown in Figure 3.5.

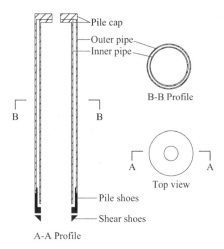

Figure 3.3 The schematic diagram of double-walled pipe pile

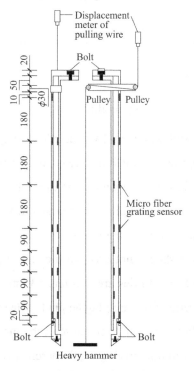

Figure 3.4 The layout of model pile sensors

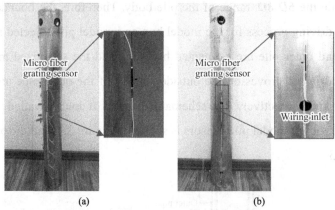

Figure 3.5 The photos of inner and outer pipe
(a) Inner pipe; (b) Outer pipe

3.1.1.3 Test program

A total of 4 groups of tests were carried out with 4 different pile tip (pile shoe) forms. The specific test scheme is shown in Table 3.2.

Test program   Table 3.2

Test number	Relative density of the sand	Loading method	Type of pile shoe
PO-1	73%	Continuous static driving	Pile shoes with camber angle of 30°
PO-2			Rectangular pile shoe
PO-3			Pile shoes with camber angle of −30°
PC			Closed flat pile shoes

3.1.2 Test results and analyses

3.1.2.1 The development law of soil plug height

The variation curve of the soil plug height versus pile driving depth for the test piles, PO-1, PO-2, PO-3, are presented in Figure 3.6. It is illustrated that the soil plug height increases with the increase of the pile driving depth, while the generation rate for the soil plug decreases gradually. For instance, the generation rate of the soil plug is the largest when the pile shoe with 30° camber angle is used, while it is the smallest when the pile shoe with −30° camber angle is used, and it shows intermediate when the pile shoe is not used. At the end of pile driving to 714mm, the soil plug heights for the PO-1, PO-2, PO-3 piles are 555mm, 535mm and 514mm, respectively. The soil plug length ratio (PLR-the ratio of the soil plug height to pile driving depth) are 0.77, 0.75 and 0.72, respectively. Figure 3.7 shows the variation curve of the incremental filling ratio (IFR-ratio of the soil plug height increment to pile driving depth increment) with the pile driving depth. As shown in Figure 3.7, the IFR value fluctuates greatly with the increase of pile driving depth. From the whole aspect, the IFR value shows a decreasing trend, and the soil plug tends to be occluded in the process of the pile driving.

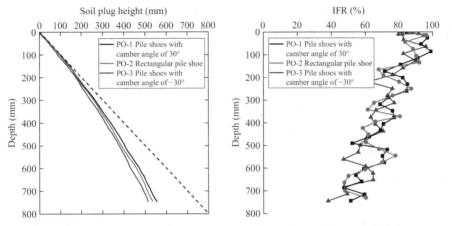

Figure 3.6 Variation curve of soil plug height with pile driving depth

Figure 3.7 Variation of IFR value of soil plug with pile driving depth

3.1.2.2 Pile driving resistance

The pile driving resistance is regarded as the macroscopic expression of coupling of various effects in the process of the pile driving. Variations of total driving resistance with the pile driving depth for the four experimental tests, are depicted in Figure 3.8. It is shown that the total driving resistance generally increases linearly with the increase of the pile driving depth, and the form of pile end has significant influence on the driving resistance of the pile. The pile driving resistance of the closed pipe pile is much greater than that of the open-ended pipe pile.

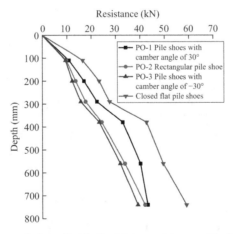

Figure 3.8 The variation of total pile driving resistance with pile driving depth

The variation of driving resistance of each part with the pile driving depth is presented in Figure 3.9. The resistance of each part increases with the increase of the pile driving depth, but the increase amplitude is quite different. Table 3.3 shows the resistance of each part and corresponding proportion. The internal and external frictional resistance of the pile account for a small proportion, while the pile wall end resistance accounts for a large proportion when the driving depth is 110mm. The proportion of the internal and external frictional resistance of the pile gradually increases when the depth of driving pile is 740mm, revealing that the internal and external frictional resistance of the pile gradually take effect in the process of driving due to the increasing contact area between the pile and surrounding soil. By comparing and analyzing the tests of PO-1, PO-2 and PO-3, it is concluded that the internal frictional resistance for the pile shoe with camber angle of 30° has the largest proportion, and the internal frictional for the pile shoe with camber angle of −30° has the smallest proportion. The rule of pile external frictional resistance is opposite, possibly due to the different contact areas. The pile end resistance of the shoeless pile is the largest, mainly because the shoeless pile has a planar annular pile end, which causes larger resistance than the inclined pile tip. It shows that the pile shoes have obvious effects on the composition of each part of pile driving resistance.

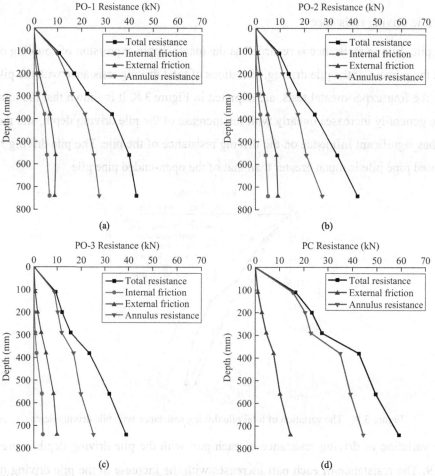

Figure 3.9   The variation curve of driving resistance of each part with the pile driving depth
(a) PO-1; (b) PO-2; (c) PO-3; (d) PC

The resistance of each part and its proportion   Table 3.3

Test number	Driving depth (mm)	Driving resistance (kN)	Pile inner frictional resistance (kN) (percentage)	Pile outer frictional resistance (kN) (percentage)	The end resistance of pile wall (kN) (percentage)
PO-1	110	10.8	0.23 (21.5%)	0.79 (7.36%)	9.77 (90.49%)
	740	45.24	6.77 (15.65%)	8.99 (20.79%)	27.48 (63.55%)
PO-2	110	9.69	0.05 (0.53%)	0.79 (8.13%)	8.85 (91.34%)
	740	42.14	5.23 (12.41%)	9.28 (22.03%)	27.63 (65.57%)
PO-3	110	9.42	0.02 (0.26%)	0.56 (5.93%)	8.84 (93.81%)
	740	39.17	3.96 (10.12%)	10.04 (25.63%)	25.16 (64.25%)
PC	110	16.54		1.15 (6.98%)	15.39 (93.02%)
	740	59.17		14.60 (24.67%)	44.57 (75.33%)

3.1.2.3　The variation rule of internal and external frictional resistance

Figure 3.10 shows the distribution of unit frictional resistance inside the pile along depth during open-ended pile driving process. Unit frictional resistance inside of the pile distributed non-uniformly along the depth at a given driving depth, and the greater the depth, the greater the frictional resistance inside of the pile. With the increase of the pile driving depth, the frictional resistance inside of the pile gradually play an increasingly important role during open-ended pile driving process. The main reasons for this phenomenon are as follows: during the pile driving process, the height of the soil plug increases continuously, and the soil plug is squeezed and compacted, which leads to the increases of its extrusion effect on the piles and the unit frictional resistance between soil plug and pile inner wall. When the pile driving depth is large, with the increase of the penetration depth, the unit frictional resistance inside the pile in the same soil layer will be weakened. When the soil plug is generated, it compresses the inner wall of pile and the unit frictional resistance of the inside of the pile is the largest. The continuous driving of the pile tip results in the rearrangement of soil particles, and the unit frictional resistance of the inside of the pile at the same depth decreases gradually. It can be concluded that there is a "side resistance degradation effect" for the frictional resistance of the inside of the open-ended pile with large diameter.

Figure 3.11 shows the variation curve for the unit frictional resistance on the outside of the pipe pile with depth during the pile driving process. It can be seen that at the same pile driving depth, the unit frictional resistance on the outside of the pipe pile also shows non-uniform distribution along the depth. The greater the depth, the greater the unit frictional resistance on the outside of the pipe pile, which is the same as the study of Iskander (2011). With the increase of the pile driving depth, the compressive effect of the pile on the surrounding soil is greater, the shear force between the outer wall of the pile and the soil is greater, and the unit frictional resistance outside the pile is greater. Under a certain pile driving depth, the unit frictional resistance outside the pile is obviously degraded. This is mainly because when the pile tip reaches a certain depth, the soil-squeezing effect is more obvious, and the unit frictional resistance on the outside of the pile reaches the maximum value. With the penetration of the pile tip, the soil particles are rearranged, and the pile side forms a "soil arch effect", which squeezes the soil. The effect is weakened. The maximum external unit frictional resistance of the closed pipe pile is 63.65kPa, which is significantly larger than that of the open pipe pile.

Comparing the Figure 3.10 and Figure 3.11, the unit frictional resistance inside the pile is smaller than that outside the pile at the beginning of the pile driving. The unit frictional resistance inside the pile increases gradually, and the gap between the unit frictional resistance inside the pile and outside the pile is reduced with the increase of the pile driving depth. It can be also seen that the bearing capacity of the soil plug plays an increasingly important role in the process of the pile driving.

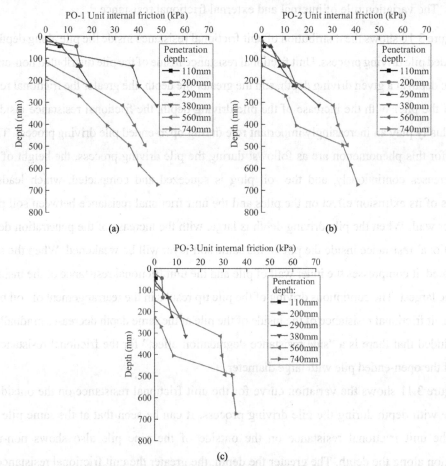

Figure 3.10 Distribution curve of unit frictional resistance on pile inner side along depth
(a) PO-1; (b) PO-2; (c) PO-3

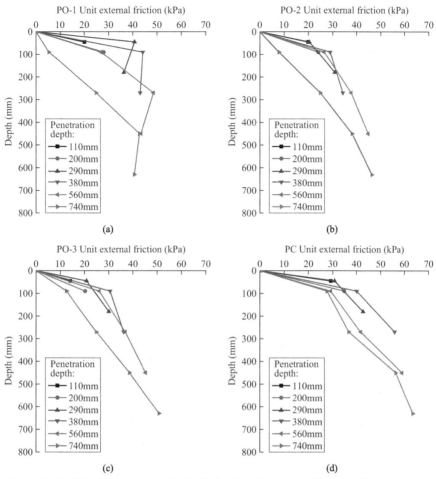

Figure 3.11 Distribution curve of unit frictional resistance on pile outer side along depth
(a) PO-1; (b) PO-2; (c) PO-3; (d) PC

3.1.2.4 Vertical displacement of the ground

In order to dynamically measure the change of soil deformation around the pile with the driving depth, a YWD-100 displacement meter was installed within the range of 0-500mm from the pile, the displacement measuring point is made of $\phi 8$ ribbed steel bars inserted into the sand at a depth of 60mm.

Figure 3.12 shows the variation of the ground movement around the pile with the driving depth. It can be seen that the ground displacement around the pile has roughly the same variation trend with the radial distance during the pile driving process, and the maximum heave-up displacement is about 30mm away from the pile. The influence of the pile driving on the ground displacement around the pile decreases with the increase of radial distance. The soil surface heave around the pile is closely related to the pile driving depth, with the increase of pile driving depth, the soil heave rate around the pile decreases gradually.

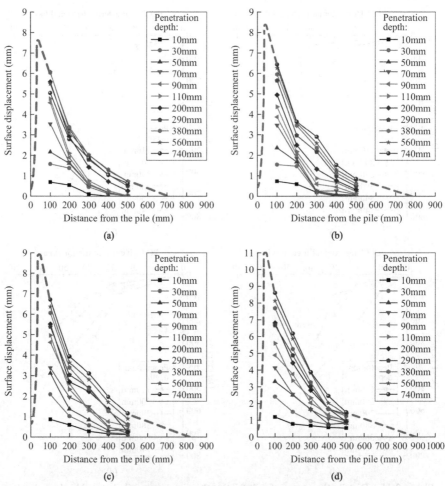

Figure 3.12 Variation curve of surface displacement with pile driving depth
(a) PO-1; (b) PO-2; (c) PO-3; (d) PC

Table 3.4 lists the vertical heave-up of the ground around the test piles in each test when the pile driving depth reaches 740mm. It can be seen that among the open-ended pipe piles, the vertical heave-up of the surrounding soil at the same radial position from the pile is the largest in the PO-3 test (shoe with camber angle at $-30°$), followed by PO-2 and PO-1. The influence of the pile body on the surrounding soil corresponds well with the height of the soil plug. When the camber angle is at 30°, the height of the soil plug is the largest, and the heave-up of the soil surface around the pile is the smallest. The heave-up of the surrounding soil surface for the closed pipe piles is the largest, and the soil squeeze effect is the most obvious. When the pile driving depth of the closed pipe pile is 740mm, the displacement of the topsoil around the pile within the range of 100-500mm is the largest, which shows that the impact of the closed pipe pile on the environment around the pile is greater than that of the open-ended pipe pile during the pile driving process.

At the end of the test, there was obvious subsidence of the soil adjacent to the pile body, which was due to the frictional dragging effect produced by the pile wall. The range of subsidence is about 30mm away from the pile body, and the amplitude is about 10mm, according to which the displacement extension line is drawn in Figure 3.12. At the same time, it can be seen that the influence range of the surrounding soil of PO-1, PO-2, PO-3 and PC test piles is about 700mm, 710mm, 810mm and 900mm. The pile diameter of the model pile in this test is 140mm, and the influence on the surrounding soil of the pile is about 5 to 7 times of the pile diameter, which is similar to the conclusion obtained by Yegian et al. (1973).

Ground heave-up around the pile at different radial distances from the pile   Table 3.4

Test number	Distance from pile (mm)				
	100	200	300	400	500
PO-1	5.05	2.79	1.86	1.04	0.63
PO-2	6.43	3.64	2.89	1.52	0.84
PO-3	6.71	3.93	3.11	1.97	1.15
PC	8.60	6.18	3.87	2.45	1.47

3.1.3 Summary

In this section, a large-scale model test is carried out on the driving process of double-walled pipe piles in sandy soil. The pile driving resistance, the unit frictional resistance of the inside and outside of the pile, the development law of the soil plug height, and the change law of the ground displacement around the pile are obtained during pile driving are obtained. The conclusions are as follows:

(1) During pipe pile driving, the soil plug height generated by the pile shoe with camber angle of 30° is the largest, and the soil plug height generated by the pile shoe with camber angle of $-30°$ is the smallest. Pile shoe has a significant effect on the generation of soil plugs. The IFR value decreases with pile driving depth, showing that the pipe pile tends to be blocked during the pile driving process.

(2) Driving resistance of the pipe pile basically increases linearly with the increase of the pile driving depth, but the form of the pile shoe affects the composition ratio of each part of the pile driving resistance. The inner frictional resistance for pile shoe with camber angle of 30° accounts for the highest proportion of the total pile driving resistance amongst the four tests, while the outer frictional resistance for pile shoe with camber angle of $-30°$ accounts for the highest proportion of the total pile driving resistance amongst the four tests.

(3) Under the same driving depth, the unit frictional resistance inside and outside the pile distributed non-uniformly along the depth, and the greater the depth is, the greater the unit

frictional resistance. When the pile driving depth is larger, there is a "lateral resistance degradation" effect on the inside and outside of the pile. The pile shoe has a significant impact on the inner and outer frictional resistance of the pile. The pile shoe with camber angle of 30° has larger unit frictional resistance than other pile shoes, and the pile shoe with camber angle of −30° has larger outer unit frictional resistance than other pile shoes.

(4) In the process of pile driving, the influence of the pile driving on the ground heave-up around the pile gradually decreases with the increase of the radial distance. When the radial position is the same, the surface heave-up around the pile gradually increases with the increase of the pile driving depth with the magnitude gradually decreasing. The influence of the pile shoe on the surrounding soil corresponds well to the height of the soil plug. The heave-up of the soil around the shoe with 30° camber angle is the smallest, while the surface heave-up around the shoe with −30° camber angle is the largest among the open-ended pipe piles. The soil surface heave-up around the closed pipe pile is the largest, and the soil squeeze effect is the most obvious. The influence range of the pipe pile on the surrounding soil is about 5-7 times the pile diameter.

3.2 DEM investigation of installation responses of jacked open-ended piles

When driving an open-ended pile into soil, part of the soil underneath is squeezed into the open end of the pile, forming a soil column, which is referred to as plugging effect. This installation effect results in distinctive behaviors of open-ended piles compared to that of the equivalent close-ended piles. The differences between them, however, vary over a wide range under various conditions (Randolph, 2003; Jardine et al., 2005). This is due to the facts that the plugging effect is closely related to a great number of factors, such as the pile characteristics (Henke & Grabe, 2013; Kumara et al., 2015; Abu-Farsakh et al., 2017), the soil conditions (Lee et al., 2003; Paik & Salgado, 2003) and the installation methods (Nicola & Randolph, 1997; De Nicola & Randolph, 1997; Paik & Salgado 2004; Henke & Grabe, 2008, 2013; Liu et al., 2012). The plugged modes (e.g. fully coring, partially and fully plugged) are usually converted from one to another even during one continue installation process, which leads to a varying relationship between the soil plug resistance and the pile annulus resistance. Meanwhile, the plugging behavior also influences the pile shaft external friction, as the flows of soil particles into the pile change the surrounding soil stresses, resulting in a somewhat lower radial stress on the pile-soil interface compared with the close-ended pile.

The application of a double-walled pile system into experiments, separating the internal and external frictions, has improved the understanding of open-ended piles (Lehane & Gavin, 2001; Paik et al., 2003; Junyoung & Sangseom, 2015; Jeong et al., 2015). A significant achievement obtained was the "wedged plug zone" where the internal friction was concentrated and balanced the soil plug resistance. Although it has been confirmed that the wedged plug zone is within only a rough range of 2-5 times of the pile diameters above the bottom, this zone exhibited remarkably different distributions in various cases (Nicola & Randolph, 1997; Paik, 1993; De Nicola & Randolph, 1997; Grabe & Heins, 2017). The similar difference was also reflected in the instabilities of predictive performances of the current design methods such as ICP-05, UWA-05 and NGI method (Jardine & Chow, 2007), which although had incorporated the plugging effect by means of adopting the macro indices of the change in plug length (e.g. plug length ratio PLR and incremental filling ratio IFR). This demonstrates that the plugging mechanism probably has not been completely understood. There are still considerable uncertainty in relation to the construction effects on the open-ended pile, particularly the stress regime inside and around the pile (from Premier geotechnical Rankine Lecture, 2016).

Understanding the micro-mechanisms is essential to interpret the macro-behavior in complex geotechnical issues (Bolton & Cheng, 2001). Advanced experimental techniques including Particle Image Velocimetry (White et al., 2003), X-ray CT (Kikuchi et al., 2008) and photographic (Zhou, 2012) have significantly advanced the understanding of the pile-soil interactions from the microscopic perspective. However, they were mainly focused on the performances within the shear zone. In fact, the systematic micro-mechanisms and their connections with macro-properties with respect to the open-ended pile installation responses, which are primarily controlled by the plugging effects, have been rarely investigated. This is probably due to the restrictions of current measuring techniques in the tests, particularly for the soil plug in the pile tube (Byrne & Houlsby, 2003).

Discrete Element Method (DEM) offers an alternative method in the detailed study of the micro-mechanics, such as inter-particle forces, movements and rotations of discrete particles. In particular, DEM-based models can be applied directly to solve larger-scale engineering problems (Cundall, 2001; Maynar & Rodríguez, 2005; Bertrand et al., 2008). However, most reported pile-related DEM studies were concentrated on the cone penetration tests (Arroyo et al., 2011; Jiang et al., 2006, 2014) and close-ended pile behaviors (Lobo-Guerrero & Vallejo, 2005; Duan & Cheng, 2016a). Only few DEM researches has considered open-ended piles in the granular soils by now (Lobo-Guerrero & Vallejo, 2007), but they limited their scopes in the crushable behavior of sand grains induced by the pile driving only. In this section, the numerical DEM method was used to reveal the comprehensive responses of the soil-pile system during the open-ended pile installation, including the micro-mechanisms and macro-behaviors both inside and outside the pile, by continuously jacking the piles with different diameters into a granular soil.

3.2.1 DEM modeling

3.2.1.1 Sample preparation

The numerical sample for this DEM modeling was prepared using the GM DEM-centrifuge method (Duan & Cheng, 2016) using the Particle Flow Code 2D (PFC^{2D}). The scaling laws commonly used in the centrifuge modeling are adopted in this study (Schofield, 1980). Rigid walls were used to model the boundary. The dimension of DEM soil model is 2.4m in width and 1.05m in depth.

At the first stage during sample generation, sand particles inside each 0.1m × 0.1m grid were generated one by one. An approximately 280 particles was created in each grid, with an initial average porosity of 0.25, and the model was cycled to reach the equilibrium state at each time. This process was repeated until all particles in all the grids were created. At the second stage, a 100g gravity force in the y direction was applied to the whole system, and the PFC model was numerically cycled again to obtain the equilibrium state. At this moment, the porosity reaches the final average porosity of 0.185. At the third stage a series of clumps were created to model the rigid monopile with a finite surface roughness.

The sand particles are made of disks with a maximum diameter of 7.05mm, a minimum diameter of 4.5mm, an average grain diameter $d_{50} = 5.85$mm and uniformity coefficient $c_u = d_{60}/d_{10} = 1.26$ (see Figure 3.13). Table 3.5 shows the input parameters used in the DEM simulations. The ratio of d_{pile}/d_{50} are around 4-16 in this model, which is close to the values suggested by Vallejo and Lobo-Guerrero (2005), ensuring the efficiency and accuracy of this numerical modeling.

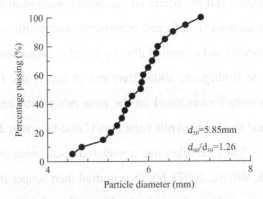

Figure 3.13   Particle size distribution used in DEM modeling (Liu et al., 2019)

Input parameters for DEM simulations   Table 3.5

Parameter	Value
Density of sand particles (kg/m^3)	2650
Density of particles for pile (kg/m^3)	66.65
Particle diameters, d (mm)	Figure 3.13
Average particle size, d_{50} (mm)	5.85
Model pile diameters, d_{pile} (mm)	22.5, 45, 90
Model pile length, L (mm)	515
Model pile wall thickness, d_{pw} (mm)	2.475
Model container width (mm)	2400
Model container depth, D (mm)	1052.3
Frictional coefficient of the particles μ (—)	0.5
Frictional coefficient of pile and walls μ (—)	0.5
Particles Young's Modulus, E_p (Pa)	4e7
Contact normal stiffness of pile & particles, k_n (N/m)	8e7
Particle stiffness ratio (k_s/k_n)	0.25
Contact normal stiffness of walls, k_n (N/m)	6e12
Initial average porosity	0.25
Final average porosity (final equilibrium)	0.185
Bulk unit weight γ bulk (kN/m^3)	2115.3

Figure 3.14(a) presents the distribution of initial void ratio in the DEM model up to 0.5m deep, equivalent to prototype depth of 50m. The average void ratio line (square marks) slightly decreases with soil depth, with a variation of 0.05. A uniformly distributed void ratio is achieved using the proposed GM method. Figure 3.14(b) compares the average lateral and vertical stress distributions along depth. The value of lateral stress coefficient K_0 is 0.65.

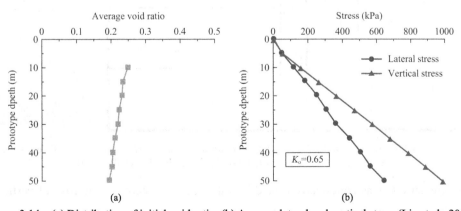

Figure 3.14   (a) Distribution of initial void ratio; (b) Average lateral and vertical stress (Liu et al., 2019)

3.2.1.2 DEM model and pile setup

All DEM analyses in this investigation were performed using an increased gravity field of 100g. Three open-ended rigid tubular piles were adopted with the same wall thickness (2.475mm) and length ($L = 0.515$m) but different outer diameters (d_{pile} = 22.5mm, 45mm and 90mm for pile P1, P2 and P3 respectively). A schematic view of the model is shown in Figure 3.15. To obtain the external and internal frictions separately, a double-walled pile system is adopted in this model, as illustrated in Figure 3.15, which is similar to those in the chamber pile tests (Lehane & Gavin, 2001; Paik et al., 2003). The pile used in this study is rigid, and made of 4940 small-sized particles with radius R_{pile} of 1.125mm, forming its three sides. The pile particles overlap each other, and the distance between the centres of two adjacent particles d_{pp} is $0.2R_{\text{pile}}$. Outer and inner rigid pile walls also are made of overlapping particles with $d_{\text{pt}} = 0.2R_{\text{pile}}$. The input density of these pile particles was scaled such that the overall pile has the weight of a steel pile. Due to the small size of particles and the short distance between every two balls, the surface of the pile is smooth enough to eliminate the occurrence of any out-of-direction shaft resistance. A view of the particle assembly with the mean diameter d_{50} = 5.85mm is also shown in the lower subset of Figure 3.15.

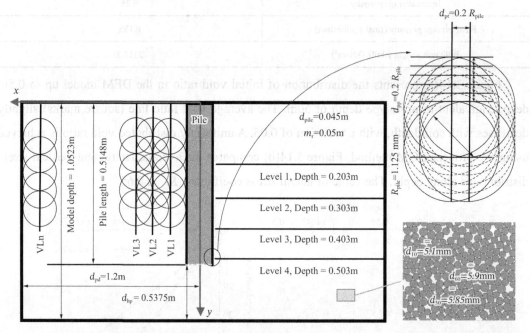

Figure 3.15 Schematic view of the PFC model; Composition of pile in PFC; A typical particle assembly at equilibrium before pile installation (not to scale) (Liu et al., 2019)

The model pile was continually jacked into the soil by a stepwise increase of vertical load until the desired depth of 0.5m was reached. Under each specific load, the system was cycled to equilibrium until the pile displacement reached its maximum value, and the following load was then applied immediately. As indicated in Figure 3.15, a series of "measurement circles" (the radius $m_r = 0.05m$) were arranged at a distance of 0.05m to each other at four levels of depths to monitor the parameters of soil elements surrounding the pile. Note that only left side of model was analyzed due to the symmetrical nature of the problem.

3.2.2 DEM simulation results

3.2.2.1 Displacements of soil around pile

As shown in the soil profile provided in Figure 3.16(a)-(c), there are clear soil convections surrounding both sides of the three piles: shallow soil adjacent to pile is pushed down by pile shaft friction; deeper soil was repelled down and outside by the pile; at the far sides of the soil tank, soils are heaved up slightly. In additon, a thin shear band forms immediately adjacent to the pile external skin after the installations, which contains mixed soil particles from the different soil layers above the corresponding level. The soils in shear band experienced large strain and downward displacement, leading to large changes in properties from virgin soils. This indicates that the pile-soil interface frictional characteristics depend not only on the local soil layer but also on the upper soil layers. The shear bandwidths for the three piles, in spite of quite different pile diameters, are almost uniform and close to the observations by Yang (2010) and White & Bolton (2004), which are in the range of 4-6 times of d_{50}. This implies that the shear bandwidths are probably controlled mainly by the grain size rather than the pile diameter. Meanwhile, it is also found that the shear bandwidth is not constant with the depth, decreasing linearly from $5d_{50}$ at ground surface to $2d_{50}$ at bottom. This phenomenon matches the observations in model pile tests (Yang, 2010). In the current simulations, it is reasonable to attribute this observation of wider shear bandwidth in the upper level to the larger shearing displacement.

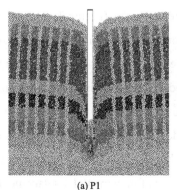
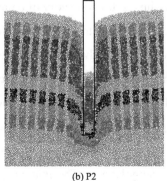
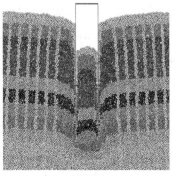

(a) P1　　　　　　　　　　　　(b) P2　　　　　　　　　　　　(c) P3

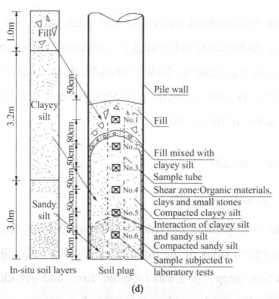

Figure 3.16 (a-c) Soil profile at pile penetration depth of 0.5m in this simulation (Liu et al., 2019); (d) Formation of the soil plug from field test (Liu et al., 2012)

The incremental displacement of soil particles, when the load on the pile increased from 40kN to 80kN, are illustration in Figure 3.17. It is clear that the soil displacement pattern follows the local shear failure mode (Figure 3.17 (d)), which has the following features: well-defined slip surfaces only below the pile, discontinuous either side. They form a triangular zone underneath the pile, where particles move downwards, and two inclined strips leaving from either side of the triangle, where particles move downwards and outwards. Comparing various pile diameters, the width of the triangular zone underneath P1 ($d_{pile} = 0.0225$m) is about 3-4 times of the pile diameter. The influence zone is reduced to only similar size to the pile diameter underneath P3 ($d_{pile} = 0.09$m). It is also observed that, in the initial load increment, from 0 to 40kN, particles in a large zone besides the pile move downwards and outwards. In the following load increments, as shown in Figure 3.17, the influence zone of pile driven reduces to limited region adjacent to pile surfaces, and the particles besides the pile mainly move downwards.

This triangular zone, however, is not stationary relative to the tip of open-ended pile during installation, which is apparent according to the trajectory of painted particles (Figure 3.16). Some particles in this triangular area are squeezed into the pile and form the soil plug, while the rest particles in the edge area flow around the pile shoulder and move into the shear zone adjacent to the pile shaft. The ratio between these two part particles is related to stress levels in the soil plug and the shear zone around pile shaft as soil particles can easily move to the zone with lower stress. This phenomenon is primarily depended on the soil plugging degree. Hence, this process also means that part of particles in the shear zone are dragged down by pile shaft from upper layers, and others are slipped out from the triangular area. The enlarged plots of incremental displacement around the pile tips in Figure 3.17 (with thick blue lines indicating the pile position under 40kN load, and thin red lines indicating the pile position under 80kN load) confirm these statements. With the movement of P1, shaft frictional forces dragged down the soil particles in the pile and underneath the pile by an amount slightly smaller than the pile displacement; as a result, the soil plug height only increases slightly during this load interval. With increasing pile diameter, soil particle displacements reduces significantly as evidenced by the lengths of displacement vectors. Therefore, soil plug height increases obviously with increasing pile diameter. It can also be observed that soil flow along pile inner surface is more significant than in the centre of pile, which leads to the "hump" at the top of soil plug as shown in Figure 3.16.

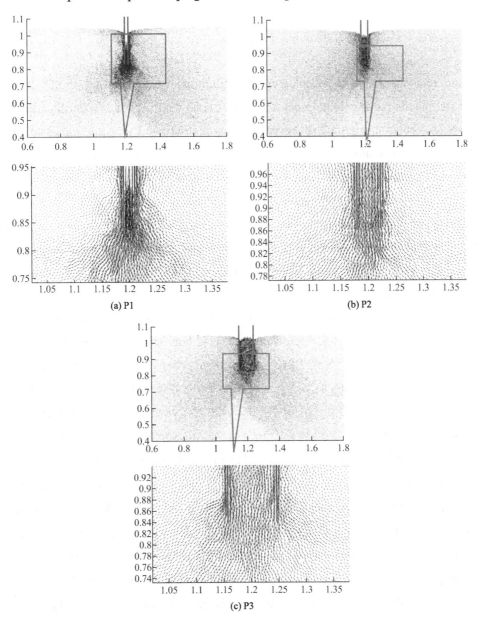

(a) P1

(b) P2

(c) P3

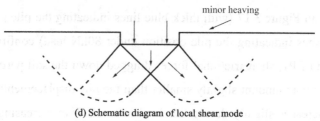

(d) Schematic diagram of local shear mode

Figure 3.17 Soil particle displacement pattern (Liu et al., 2019)

3.2.2.2 Plugging behavior

To quantify the plugging behavior, plug length ratio (PLR) and incremental fill ratio (IFR) are defined in literature (Liu et al., 2012) as

$$\text{PLR} = \frac{H}{L} \quad (3.1)$$

$$\text{IFR} = \frac{dH}{dL} \times 100(\%) \quad (3.2)$$

where H is the length of soil plug corresponding to an embedded length of pile L, while dH and dL are their instant increments. The variations in the PLR and IFR during installation of three piles are shown in Figure 3.18, revealing clear different soil plugging trends for the three piles. It is reasonable to attribute these differences to the pile sectional dimension as all the other parameters are set to be the same in this modeling. For the slimmest pile P1, fully plugged mode (IFR = 0) is achieved as pile tip penetrates to the depth of 0.2m, and then it is transferred to partially plugged mode (0 < IFR < 1) at penetrate depth of 0.3m, but is fully plugged again at penetrate depth of 0.4m. In comparison, Pile P2 produces lower degree of soil pugging, where fully plugged mode is found only at the depth of 0.25m and the final stage. Pile P3 with the largest diameter, however, does not achieve fully plugged mode during almost the entire installation process until the final penetration depth at 0.5m. Evidently the plugging effect of open-ended piles decreases with an increase of the pile diameter. This is consistent with previous findings (e.g., Gudavalli et al., 2013; Jeong et al., 2015), which demonstrated that piles with larger diameter tended to have a partially-plugged penetrate mode. These different plugging trends lead to the different final lengths of soil plugs for P1, P2 and P3, which are 0.11m, 0.24m and 0.35m respectively, equivalent to 4.8, 5.3 and 3.9 times of the corresponding pile diameters.

Compared with PLR, IFR is more preferred in the design methods such as UWA-05 approach and other researches (Paik & Lee, 1993; Paik & Salgado, 2003) to quantify the plugging degree. However, this "moving" parameter is hardly recorded in real applications, particularly in the offshore environment. Figure 3.18 shows that IFR and the corresponding PLR values exhibit similar trends, decreasing roughly with penetration depth, despite IFR being more sensitive to the installation process. It is therefore meaningful to build the relationship of IFR to PLR, as PLR is easily measured at the end of installation process. Figure 3.19 plots the IFR versus PLR obtained from this modeling, as well as some other published data: Szechy (1959), Klos & Tejchman (1977), Brucy et al. (1991), Paik et al. (2002), Paik & Salgado (2003) and Jeong et al. (2015). The current modeling data generally agree with these published data. Despite the scattering of the data, which is probably a result of the differences in soil properties and installation methods, a linear correlation between IFR and PLR can be observed for the entire set of data and can be expressed as follows:

$$IFR(\%) = 106.14 \times PLR - 16.44 \tag{3.3}$$

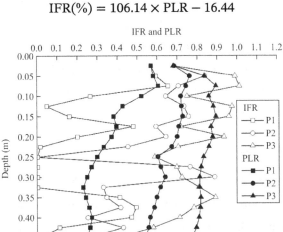

Figure 3.18 Development of IFR and PLR of piles P1, P2, and P3 during installation (Liu et al., 2019)

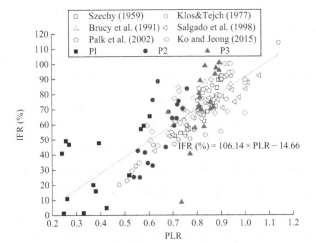

Figure 3.19 Relationship between IFR and PLR (Liu et al., 2019)

The layers of soil plug follow the virgin deposit in sequence, as shown in Figure 3.16(a)-(c), but they are clearly not flat for all three piles. The soil plug in pile P2, for instance, exhibits convexity at the upper part but has concave shapes at the lower part. This shows "initiative arch" forming in the early jacking process and followed by "passive arch" in the later jacking process. This is consistent with the plugging trend that pile P2 tends to be plugged in the later installation. This soil-arching behavior is valiated by the full-scale tests (Liu et al., 2012) (Figure 3.16(d)) and model tests (Zhou,

2012). It is thus reasonable to believe that the formation of passive arch significantly enhances the resistance of soil plug, therefore, soil plug is being fully plugged simultaneously.

In practice, the pile with zero value of IFR is considered to be fully plugged, as such condition can be conveniently identified in the physical tests by mornitoring the soil plug length. However, the comparison between Figure 3.16 and Figure 3.18 shows that this "apparently plugged" is not the "truly plugged" that performs completely like a close-ended pile. Take P2 as instance, although the soil plug length is constant in the later 0.25m penetration, the soils beneath the pile tip still further intrude into pipe, leading to the soils inside the pipe being further compacted. This process mobilizes a different base resistance compared to close-ended pile due to different movements of soil particles in this area. The traditional load transfer model at pile base obtained from close-ended pile tests, therefore, should not be applied directly to open-ended pile under fully plugged mode.

3.2.2.3 Installation resistance

Two-dimensional model can be regarded as a thin slice in the middle of the pile, so that the forces in this slice with unit width are used to compare the installation resistances for the three piles. The overall installation resistance Q_p, as well as its three components including the external friction Q_{os}, the internal friction Q_{is} (equal to the plug resistance Q_{plug}) and the annulus resistance Q_{ann}, are plotted against the penetration depth in Figure 3.20(a), (b) and (c) for Pile P1, P2 and P3 respectively. It is clear for all three piles that the three resistant components are almost identical in magnitude during the initial shallow penetration (approximately 0-0.25m), but exhibit differences during the deep penetration. For Pile P1, the external friction exceeds internal friction from the penetration depth of 0.325m, as the internal friction drops abruptly while the external friction jumps at that time. This is consistent with the IFR trend in Figure 3.18, showing Pile P1 becomes unplugged from this depth. This relationship can also be found in Figure 3.20(b) for Pile P2, where the internal friction decreases evidently as the pile is transformed from plugged to unplugged (0.25m), but the external friction is the reverse. Even so, the internal friction of Pile P2 still remains at similar values to the external friction in the deeper penetration, and both of them are larger than the corresponding annulus resistance. In comparison, the internal friction of Pile P3 increases sharply in the latter half phase, which reaches almost twice of the sum of the external friction and annulus resistance at the end of installation.

The final total resistances of the three model piles are 211kN, 295kN and 297kN respectively. Evidently the pile base (sum of the soil plug and pile annulus) contributes to the majority of them, and this ratio increases significantly with an increase in pile diameter. The unit plug resistances (q_{plg}) and unite annulus resistances (q_{ann}) (resistance per unit pile length) during installation for three piles are plotted in Figure 3.21. Regardless of fluctuations during the installation process, the unit annulus resistances for all three piles exhibit almost the same trends with penetration, despite different plugging degrees. This implies that the unit annulus resistance (q_{ann}) is independent on pile scale and plugging mode, similar to the observations in some previous experiments (e.g., Lehane & Gavin, 2001; Liu et al., 2012).

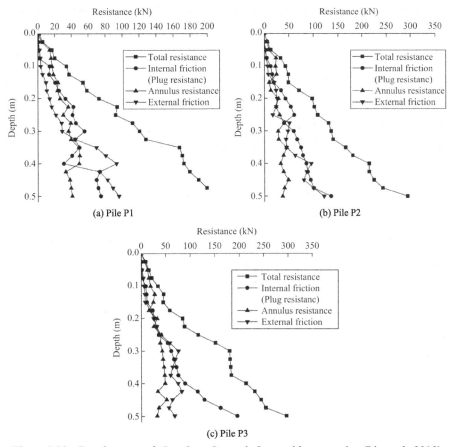

Figure 3.20　Development of Q_p, Q_{os}, Q_{is} and Q_{ann} with penetration (Liu et al., 2019)

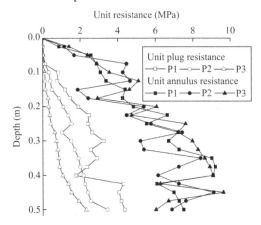

Figure 3.21　Development of q_{plg} and q_{ann} with penetration (Liu et al., 2019)

However, the unit plug resistances are evidently different for three piles. During penetration, pile P1 with highest plugging degree (lowest IFR value) among these three piles has the largest unit plug resistance. This is further validated by the similar relationship observed by Lehane and Gavin (2001) and Doherty et al. (2010) that the ratio of unit plug resistance to CPT-q_c decreases with increasing IFR value. At final penetration, all three piles reach the fully plugged mode and behavior resemble close-ended piles, but the unit plug resistances for P1, P2 and P3 are still inconsistent, which are 72%, 100% and 159% less than the unit annulus resistance respectively. It is reasonable to attribute this difference to the pile scale, as the ICP design method (Jardine et al, 2005) suggests that the unit tip resistance fall logarithmically with increasing pile diameter for both closed and open ended pile.

The external friction for three piles, as shown in Figure 3.21, increases much more gently and even experience some drops during driving. This implies that the local external friction may possibly decrease with continued shearing at the pile-soil interface during further installation. This is testified by the profile of unit external friction at the different installation depths in Figure 3.22, showing evidence for degradation in unit external friction at a given depth as the distance h to the pile tip increases. This phenomenon is referred to as "friction fatigue" (Heerema, 1978) or the "h/R effect" (Bond & Jardine, 1991). It is widely accepted that the decay is primarily attributed to the reduction of horizontal effective stress acting on the pile shaft, which is induced by the thickness reduction of shear band (Randolph, 2003).

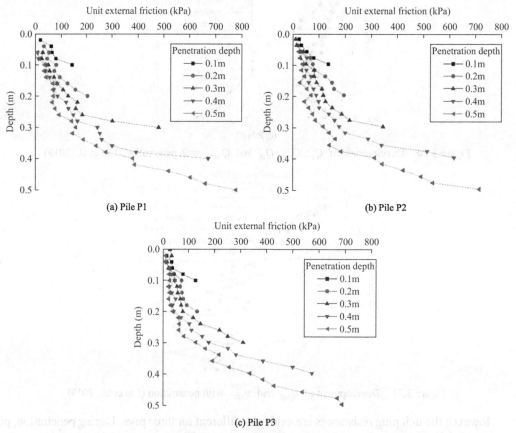

Figure 3.22 Unit external friction at the different installation depths (Liu et al., 2019)

A degradation factor, first proposed by Randolph et al. (1994), has been generally adopted to quantify the decay rate of pile-soil interface friction. Two different values of 0.38 and 0.5 are adopted respectively in the ICP method (Jardine et al., 2005) and the UWA method (Lehane et al., 2005) to design steel tabular piles in sand. The obtained reduction in local external shaft friction(f_{os}) at four given depths for the three piles in the current simulations, normalized by the initial "unfatigued" friction(f_{os0}) at the pile tip during the installation, as a function of normalized distance from the pile base h/D, is plotted in Figure 3.23. A regression analysis was performed on the data, leading to the following equation with the degradation factor of 0.458, which is close to the value of 0.41 obtained from full-scale driven piles in uniform sand by Flynn and Bryan (2015).

$$f_{os} = f_{os0}(h/D) - 0.458 \qquad (3.4)$$

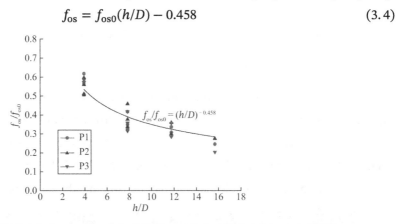

Figure 3.23 Reduction in local external shaft friction with installation (Liu et al., 2019)

The unit internal frictions in the pile-plug interface developed in the installation are also plotted in the Figure 3.24. It is obvious that the internal friction is much higher than the outer resistance at the same level. The final unit internal friction just above the pile base are 1380kPa, 1180kPa and 1000kPa, which are 50%, 48%, 36% larger than the unit external friction at the same level. Although the internal frictions are mobilized along the whole soil plug, most of them are concentrated within the zone of 2-3 times of d_{pile} above the pile base. This is consistent with the observations that the resistances are developed only in the lower portion of the soil column under vertical loading (De Nicola & Randolph, 1997; Lehane & Gavin, 2001).

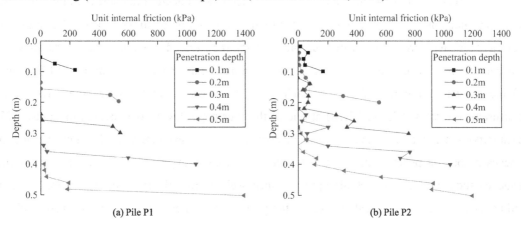

(a) Pile P1   (b) Pile P2

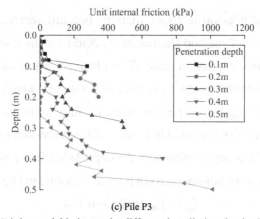

(c) Pile P3

Figure 3.24   Unit internal friction at the different installation depths (Liu et al., 2019)

3.2.2.4   Contact force network

The contact force network surrounding each pile was visualized in Figure 3.25. Each line segment connects the centroids of two contacting particles with line thickness proportional to the force magnitude. The strong contact force network within the pile transmits up further in the smaller diameter (P1), which prevents the uprising of soils. However in P3, the strong contact force network only appears at the tip of the pile and does not transmit up, therefore soils rise up more easily.

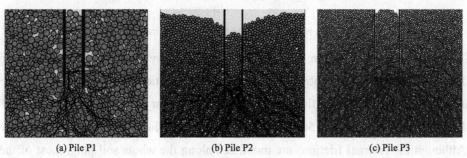

(a) Pile P1   (b) Pile P2   (c) Pile P3

Figure 3.25   Contact force network surrounding the pile (Liu et al., 2019)

3.2.2.5   Stresses in the soil mass

The contact force network could show the force/stress concentration surrounding a pile at a particular moment, but it could not depict the temporal evolution of stresses easily. Therefore the historical output of stresses in the measurement circles as described in Section 3.2.1.2 are analyzed. The evolutions of horizontal stresses in the surrounding soil during the installations of P1, P2 and P3 are shown in Figures 3.26-3.28. Each subfigure shows lateral stresses measured in 10 measurement circles (VL1 to VL10) at one of the four depth levels as indicated in Figure 3.27. Each curve represents the lateral stresses measured in one measurement circle when the pile tip reaches a specific penetration depth. The general variation of lateral stress displays a similar tendency and matches the penetration process fairly well. As the pile tip penetrates through the soil and advanced down, the lateral stresses at the pile tip level increases sharply as a result of soil compaction, conforming to Figure 3.28 which shows that a penetrating pile base resembles a spherical cavity expansion. In the close regions at a specific level around the pile shaft, the lateral stresses climb to the peak values as the pile tip almost penetrates to this level, and then fall sharply as the penetration continues. This relaxation in radial stresses with increasing h/R has also been observed in field and model tests (e.g., Liu et al., 2012; Yang et al., 2014).

As the radial distance from the pile axis increases, this "bend" of the curve becomes less remarkable. Beyond 6 times of the pile diameter, the lateral stress builds up uniformly during the whole penetration process. This indicates that the zone of the major stress disturbance, showing significant granular behavior, has an approximate extent of 6 times of pile diameter. The greatest increase of induced lateral stress registered at level 2 (depth-0.2m) is 480Pa, which is 70% and 22% greater than those at level 3 and level 4 respectively. This observation indicates that the magnitude of the induced lateral stress increase depends not only on the radial distance but also on the embedment depth or the overburden pressure.

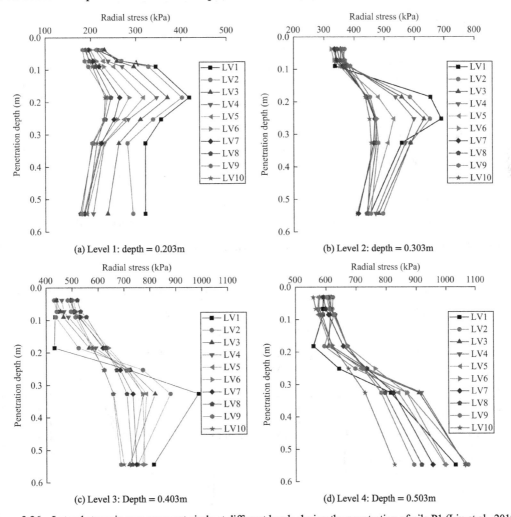

Figure 3.26 Lateral stress in measurement circle at different levels during the penetration of pile P1 (Liu et al., 2019)

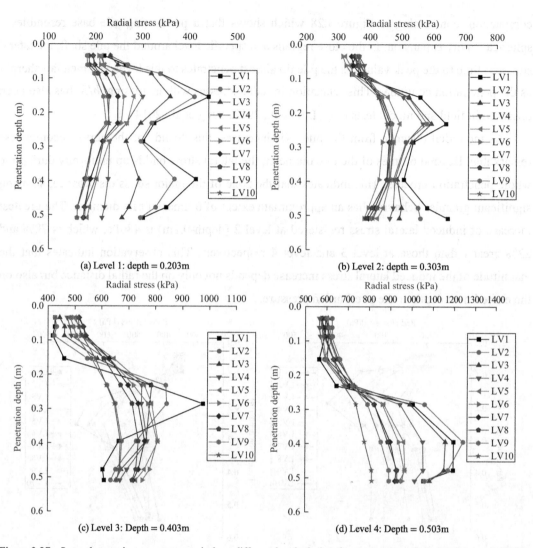

Figure 3.27 Lateral stress in measurement circle at different levels during the penetration of pile P2 (Liu et al., 2019)

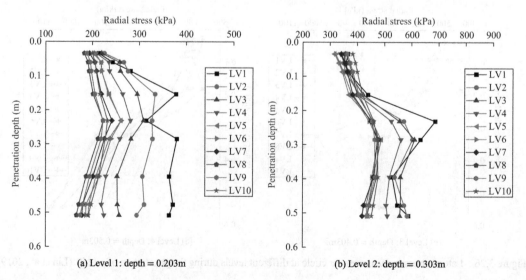

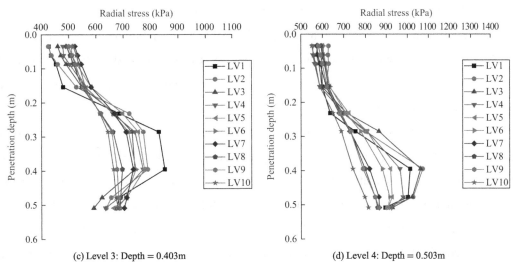

(c) Level 3: Depth = 0.403m   (d) Level 4: Depth = 0.503m

Figure 3.28 Lateral stress in measurement circle at different levels during the penetration of pile P3 (Liu et al., 2019)

3.2.3 Summary

In this section, a series of two-dimentional discrete element modeling of open-ended pile installation tests was described. The installation of this type of pile with various diameters in granular deposit are examined in detail, including the particle displacement pattern, behavior of soil plugging, the jacking resistances in the pile, and the distribution and variations of horizontal stress in soils. The main findings can be summarized as follows:

- The particle displacement pattern shows that the shear band width along pile shaft decreases from ground surface to the pile end. The shear band width is mainly governed by d_{50} and embedded depth, but independent on the pile diameter.

- The soils underneath the pile follow the local shear failure mode, where triangular slip surfaces are formed, with part of particles squeezed into the pile forming soil plug and the rest pushed downwards and outwards, forming cavity expansion.

- The instantly fully plugged mode is achieved earlier and more easily in the slimmer pile. However, the fully plugged mode and the partially plugged mode continuously exchange during pile installation, therefore the base resistance of a fully plugged open-ended pile is still lower than that of the close-ended pile.

- The pile base resistance (sum of the soil plug resistance and pile annulus resistance) contributes to the majority of pile resistance, and this ratio increases with increasing pile diameter. The unit annulus resistance is independent on pile diameter, while the unit plug resistance increases with increasing pile diameter. The plug resistance is only concentrated within the zone of 2-3 times of pile diameter above the pile base.

- The unit external friction at a specific depth decreases in pile installation, which is mainly due to the reduction of horizontal effective stress acting on the pile shaft.
- Pile installation causes increases in the average stresses surrounding the pile, while this increase decays quickly with increasing distance to pile shaft, with the major influence zone concentrated within $6d_{pile}$.

It is fully recognized that the ratios of pile diameter to soil particle mean diameter in current DEM models are around the lowest value as suggested by other researchers (in particular P1) due to the limitation of computational power. It may cause boundary effects. Moreover, the soil particles are modelled as in spherical shape and only a thin slice along the middle of the pile is considered as a two-dimentional model. These simplifications may not capture all features of soil-pile interactions pricisely. However, all simulations are carefully validated with previous laboratory tests and field tests. It has demonstrated that the current simulations qualitatively captured the trend for soil flow, soil plug, pile resistance and stress distributions. The current simulations even quantitatively captured the relationship between IFR and PLR and the shaft friction reduction. But it needs to be noted that, due to the different model scale adopted in the current models, the quantities of stress, shaft resistance etc. could not applied directly into practice without further validations and calibrations.

References

Abu-Farsakh M Y, Haque M N, Tsai C. A full-scale field study for performance evaluation of axially loaded large-diameter cylinder piles with pipe piles and PSC piles. Acta Geotechnica, 2017, 12(4): 753-77.

Arroyo M, Butlanska J, Gens A, et al. Cone penetration tests in a virtual calibration chamber. Géotechnique, 2011, 61(6): 525-531.

Bertrand D, Nicot F, Gotteland P, et al. Discrete element method (DEM) numerical modeling of double-twisted hexagonal mesh. Canadian Geotechnical Journal, 2008, 45(8): 1104-1117.

Bolton M D, Cheng Y P. Micro-geomechanics. Constitutive and centrifuge modelling: Two extremes, SM Springman, 2002: 59-74.

Bond A J, Jardine R J. Effects of installing displacement piles in a high OCR clay. Géotechnique, 1991, 41(3): 341-363.

Brucy F, Meunier J, Nauroy J F. Behavior of pile plug in sandy soils during and after driving. In: Proceedings of 23rd Annual Offshore Technology Conference. Houston, 1991: 145-154.

Byrne B W, Houlsby G T. Foundations for offshore wind turbines. Philosophical Transactions of the Royal Society of London, Series A (Mathematical, Physical and Engineering Sciences), 2003, 361(1831): 2909-2930.

Choi Y, O'Neill M W. Soil plugging and relaxation in pipe pile during earthquake motion. Journal of Geotechnical and Geoenvironmental Engineering, 1997, 123(10): 975-982.

Cundall P A. A discontinuous future for numerical modelling in geomechanics? Proceedings of the Institution of Civil Engineers-Geotechnical Engineering, 2001, 149(1): 41-47.

Cundall P A. A computer model for simulating progressive large-scale movement in blocky rock systems. Symp. ISRM, Nancy, France, Proc, 1971: 129-136.

De Nicola A, Randolph M F. The plugging behavior of driven and jacked piles in sand. Géotechnique, 1997, 47(4): 841-856.

Doherty P, Gavin K, Gallagher D. Field investigation of base resistance of pipe piles in clay. Geotechnical Engineering, 2010, 163(1): 13-22.

Duan N, Cheng Y P. A modified method of generating specimens for a 2D DEM centrifuge model. GEO-CHICAGO 2016: Sustainability, Energy, and the Geoenvironment, 2016: 610-620.

Duan N, Cheng Y P. A Modified Method of Generating Specimens for 2D DEM Centrifuge Model. GEO-CHICAGO, 2016: 610-620.

Duan N, Cheng Y P. Discrete Element Method Centrifuge Model of Monopile under Cyclic Lateral Loads. International Journal of Environmental, Chemical, Ecological, Geological and Geophysical Engineering, 2016, 10(2): 189-194.

Flynn K N, Mccabe B A. Shaft resistance of driven cast-in-situ piles in sand. Canadian Geotechnical Journal, 2016, 53(1): 49-59.

Flynn K N, McCabe B A. Shaft resistance of driven cast-in-situ piles in sand. Canadian Geotechnical Journal, 2015, 53(1): 49-59.

Gavin K G, Lehane B M. The shaft capacity of pipe piles in sand. Canadian Geotechnical Journal, 2003, 40(1): 36-45.

Grabe J, Heins E. Coupled deformation-seepage analysis of dynamic capacity tests on open-ended piles in saturated sand. Acta Geotechnica, 2017, 12(1): 211-223.

Gudavalli S R, Safaqah O, Seo H. Effect of soil plugging on axial capacity of open-ended pipe piles in sands. Proceedings of the 18th International Conference on Soil Mechanics and Geotechnical Engineering, Paris, 2013: 1487-1490.

Heerema E P. Predicting pile driveability: Heather as an illustration of the "friction fatigue" theory. Paper presented at the SPE European Petroleum Conference. Society of Petroleum Engineers, 1978.

Henke S, Grabe J. Field measurements regarding the influence of the installation method on soil plugging in tubular piles. Acta Geotechnica, 2013, 8(3): 335-352.

Henke S, Grabe J. Numerical investigation of soil plugging inside open-ended piles with respect to the installation method. Acta Geotechnica, 2008, 3(3): 215-223.

Iskander M. Behavior of pipe piles in sand: plugging & pore water pressure generation during installation and loading. Berlin: Springer-Verlag Berlin Heidelberg, 2011.

Jardine R J, Chow F C. Some recent developments in offshore pile design. Offshore Site Investigation and geotechnics, Confronting New Challenges and Sharing Knowledge. Society of Underwater Technology, 2007.

Jardine R, Chow F, Overy R, et al. ICP Design Methods for Driven Piles in Sands and Clays. Thomas Telford, London, 2005.

Jardine R. Premier geotechnical Rankine Lecture, Imperial College, London, 2016.

Jeong S, Ko J, Won J, et al. Bearing capacity analysis of open-ended piles considering the degree of soil plugging. Soils and Foundations, 2015, 55(5): 1001-1014.

Jiang M J, Yu H S, Harris D. Discrete element modelling of deep penetration in granular soils. International Journal for Numerical and Analytical Methods in Geomechanics, 2006, 30(4): 335-361.

Jiang M, Dai Y, Cui L, et al. Investigating mechanism of inclined CPT in granular ground using DEM. Granular Matter, 2014, 16(5): 785-796.

Kikuchi Y, Sato T, Morikawa Y, et al. Visualization of Plugging Phenomena in Vertically Loaded Open-Ended Piles. GeoCongress, 2008: 118-124.

Klos J, Tejchman A. Analysis of behavior of tubular piles in subsoil. In Proceedings of the 9th International Conference on Soil Mechanics and Foundation Engineering. Tokyo, Japan, 1977: 605-608.

Ko J, Jeong S. Plugging effect of open-ended piles in sandy soil. Canadian Geotechnical Journal, 2014, 52(5): 535-547.

Kumara J J, Kikuchi Y, Kurashina T. Effective Length of the Soil Plug of Inner-Sleeved Open-Ended Piles in Sand. Journal of GeoEngineering, 2015, 10(3): 75-82.

Lee J, Salgado R, Paik K. Estimation of the load capacity of pipe piles in sand based on CPT results. Journal of Geotechnical and Geoenvironmental Engineering, 2003, 129(5): 391-403.

Lehane B M, Gavin K G. Base resistance of jacked pipe piles in sand. Journal of Geotechnical and Geoenvironmental Engineering, 2001, 127(6): 473-480.

Liu J, Duan N, Cui L, et al. DEM investigation of installation responses of jacked open-ended piles. Acta Geotechnica. 2019, 14(6): 1805-1819.

Liu J W, Zhang Z M, Yu F, et al. Case history of installing instrumented jacked open-ended piles. Journal of Geotechnical and Geoenvironmental Engineering, 2012, 138(7): 810-820.

Lobo-Guerrero S, Vallejo L E Influence of pile shape and pile interaction on the crushable behavior of granular materials around driven piles: DEM analyses. Granular Matter, 2007, 9(3-4): 241-250.

Lobo-Guerrero S, Vallejo L E. DEM analysis of crushing around driven piles in granular materials. Géotechnique, 2005, 55(8): 617-623.

Maynar M J, Rodríguez L E. Discrete Numerical Model for Analysis of Earth Pressure Balance Tunnel Excavation. Journal of Geotechnical and Geoenvironmental Engineering, 2005, 131(10): 1234-1242.

Paik K H, Lee S R. Behavior of soil plugs in open-ended model piles driven into sands. Marine Georesources & Geotechnology, 1993, 11(4): 353-373.

Paik K H, Lee S R. Behavior of soil plugs in open-ended model piles driven into sands, Marine Georesources and Geotechnology, 1993, 11(4): 353-373.

Paik K, Salgado R, Lee J, et al. Behavior of open and closed-ended piles driven into sands. Journal of Geotechnical & Geoenvironmental Engineering, 2003, 129(4): 296-306.

Paik K, Salgado R, Lee J, et al. Behavior of Open-and Closed-Ended Piles Driven Into sands. Journal of Geotechnical and Geoenvironmental Engineering, 2003, 129(4): 296-306.

Paik K, Salgado R, Lee J, Kim B. Behavior of Open-and Closed-Ended Piles Driven Into sands. Journal of Geotechnical and Geoenvironmental Engineering, 2003, 129(4): 296-306.

Paik K, Salgado R. Determination of bearing capacity of open-ended piles in sand. Journal of Geotechnical and Geoenvironmental Engineering, 2003, 129(1): 46-57.

Paik K, Salgado R. Effect of pile installation method on pipe pile behavior in sands. Geotechnical Testing Journal, 2003, 27(1): 78-88.

Randolph M F, Dolwin J, Beck R. Design of driven piles in sand. Géotechnique, 1994, 44(3): 427-448.

Randolph M F. Science and empiricism in pile foundation design. Géotechnique, 2003, 53(10): 847-876.

Rao S N, Ramakrishna V G S T, Raju G B. Behavior of pile-supported dolphins in marine clay under lateral loading. Journal of Geotechnical Engineering, 1996, 122(8): 607-612.

Schofield A N. Cambridge Geotechnical Centrifuge Operations. Géotechnique, 1980, 30(3): 227-268.

Szechy C H. Tests with tubular piles. Acta Technica, Hungarian Academy of Science, 1959,

24(1-2): 181-219.

Vallejo L, Lobo-Guerrero. DEM analysis of crushing around driven piles in granular materials. Géotechnique, 2005, 55: 617-623.

White D J, Bolton M D. Displacement and strain paths during plane-strain model pile installation. Géotechnique, 2004, 54(6): 375-397.

White D J, Take W A, Bolton M D. Soil deformation measurement using particle image velocimetry (PIV) and photogrammetry. Géotechnique, 2003, 53(7): 619-631.

Yang Z X, Jardine R J, Zhu B T, et al. Sand grain crushing and interface shearing during displacement pile installation in sand. Géotechnique, 2010, 60(6): 469-482.

Yang Z X, Jardine R J, Zhu B T, et al. Stresses developed around displacement piles penetration in sand. Journal of Geotechnical and Geoenvironmental Engineering, 2014, 140(3): 04013027.

Yang Z X, Jardine R J, Zhu B T, et al. Stresses Developed around Displacement Piles Penetration in Sand. Journal of Geotechnical and Geoenvironmental Engineering, 2014, 140(3): 04013027.

Yegian M, Wright S G. Lateral soil resistance displacement relationships for pile foundation in soft clays. Offshore Technology Conference, Houston, 1973: 893.

Zhou J, Chen X L, Wang G Y, et al. Experimental and Numerical Analysis of Open-ended Pipe Piles During Jacking into Sand. Journal of Tongji University (Natural Science), 2012, 40(2): 173-178.

Chapter 4 Bearing characteristics of monopile foundations for offshore wind turbine

4.1 Failure modes of rigid monopile foundation in soft clay under multidirectional loads

4.1.1 Finite element numerical modeling

4.1.1.1 Numerical model of monopile foundation in soft soil

The NREL5MW OWT was taken as an example, and the steel monopile foundation in soft clay seabed was used in this project. The numerical model of the monopile foundation in soft clay seabed was established using ABAQUS based on the in-situ data. It included two parts, i.e., the large-diameter monopile foundation and soft clay seabed. The monopile foundation had a diameter of 5m and an embedded length of 30m. The diameter and height of soft soil seabed were set as 140m and 100m to reduce the boundary effect. The numerical model bottom boundary was restricted in x, y and z-axis directions, and the surrounding boundaries were restricted in x and y-axis directions, while the upper boundary had no restriction. The constitutive model of the steel monopile foundation was linear elastic, with the Poisson's ratio = 0.24 and Elastic modulus = 2.1×10^8 GPa. The *M-C* constitutive model was used to describe the soft soil, and model parameters of soft clay were in accord with experimental results determined in the laboratory. The internal friction angle, cohesion and elasticity modulus were determined as 32.5°, 10.2kPa and 10.4MPa.

The contact status of the monopile-seabed interface was small sliding model, and the surface-to-surface master/slave contact pair formulation was set as the contact relation between monopile and soil. The monopile surfaces were defined as the master surfaces, whereas the soil surfaces around the monopile were defined as slave surfaces. The elements of pile and soil were both C3D8R. The finite element numerical model of the large-diameter monopile foundation for the offshore wind turbine in soft soil seabed is shown in Figure 4.1.

4.1.1.2 Model verification

Matlock (1970) carried out horizontal static and cyclic load tests of the steel-pipe pile in underwater soft clay near lake Austin, Texas, USA. The steel pile foundation diameter in this test was 324mm, and the thickness was 12.7mm, while the length was 12.81m. The bending stiffness (EI) of the steel pile was 31.28MN·m^2 and the equivalent stiffness was 5.78×10^4MPa. The position of the external load imposed on the pile was 0.0635m away from the soil surface, and the displacement of pile head and strain of pile body were measured by displacement meters and strain gages arranged on the top and body of the pile. According to the field data of the pile foundation and soil parameters, a three-dimensional finite element model was established based on the large-scale finite element analysis software ABAQUS. The Mohr-Coulomb model was used for the soil, and the linear elastic model was adopted for the steel pile. The calculated parameters of the steel pile foundation and soil are shown in Table 4.1. The pile-soil interaction was simulated by the Coulomb friction model with a friction coefficient of 0.3, and the continuum element type (C3D8R) was used to model the soil and pile. The stepwise horizontal load was applied to the pile, and the relationship between the horizontal load and displacement was plotted. The comparison of numerical results and measured data in-situ was shown in Figure 4.2.

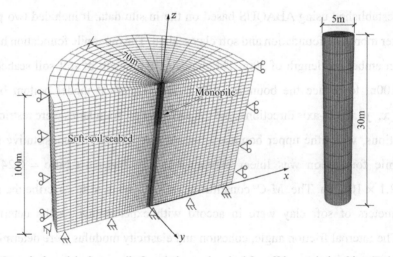

Figure 4.1　Numerical model of monopile foundation and seabed for offshore wind turbine (Dai et al., 2022)

Parameters of pile and soil in numerical calculation　　　　Table 4.1

Parameter	E (MPa)	v	c (kPa)	φ (°)
Steel pile	57800	0.2	—	—
Soft soil	2.8	0.3	26	23

It can be seen from Figure 4.2 that the relationship between the horizontal loads and the horizontal displacements of the pile head in numerical calculation and the field test were in good agreement. The horizontal movement error of the numerical calculation and the field test was small. Therefore, the finite element method could well reproduce the horizontal loading behavior of the monopile foundation in soft soil.

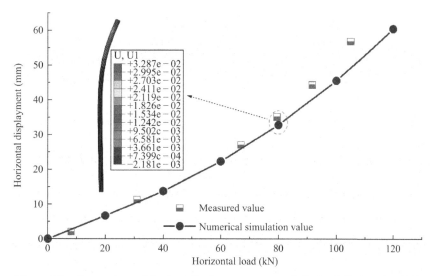

Figure 4.2 Comparison of the load-displacement curve of pile head (Dai et al., 2022)

4.1.1.3 Loading methods

There are two loading ways to choose in the pre-processing of finite element numerical model calculation, i.e., the loading controlling method and the displacement controlling method. The displacement controlling method was adopted in Section 4.1.2, and the stress and displacement of the pile top were extracted. So, the loads and displacements relationship between the monopile foundation under unidirectional loads was obtained. For example, the ultimate vertical or horizontal displacement was applied to the pile head. Then the relationship between the vertical or horizontal displacement and the vertical or horizontal load could be achieved. For gaining the failure envelopes of the monopile foundation under multi-directional loads, the fixed displacement ratio loading method was used in Section 4.1.3. Displacements in the direction i and j were implemented on the geometric center of monopile head simultaneously, and the displacement values in these two directions were respectively defined as u_i and u_j. The specific implementation steps of the fixed displacement ratio loading method were as follows:

(1) A fixed proportion of u_i/u_j was imposed on the pile head firstly.

(2) Displacements in both directions were increased until reaching the failure stage, but the du_i/du_j ratio was kept constant.

(3) Different proportions of du_i/du_j were calculated in sequence based on those above steps, and many corresponding failure points were determined.

(4) Those failure points under different proportions of du_i/du_j were connected by a line, so the failure envelope in a load space could be obtained. The analysis and calculation method of the fixed displacement ratio loading method is shown in Figure 4.3.

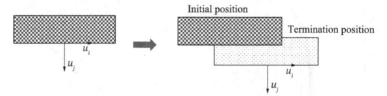

Figure 4.3 Schematic diagram of fixed displacement ratio loading method (Dai et al., 2022)

4.1.2 Failure mode of monopile foundation under unidirectional loads

The structures (foundation, tower and blades) of OWT are subjected to long-term coupling actions of combining wind, wave and current load. There are different failure modes of the monopile foundation of OWT under different unidirectional loads. This section applied the displacement-controlled method to investigate the large-diameter monopile foundation failure mode for OWT. The unidirectional horizontal, vertical or bending moment load was ordinally imposed on the monopile foundation pile head. The ultimate bearing capacity and failure mode of the monopile foundation were confirmed subsequently. The horizontal load, vertical load, and bending moment were expressed by symbol H, V, M, respectively. In contrast, the ultimate horizontal load, vertical load, and bending moment were represented by H_{ult}, V_{ult} and M_{ult}.

4.1.2.1 Failure mode of monopile foundation under unidirectional horizontal load

Different levels of horizontal displacements were imposed on the geometric center of the top of the monopile foundation. Each stage of horizontal displacements and the corresponding horizontal loads were extracted, and the horizontal load-displacement response of the monopile foundation of OWT was obtained, as shown in Figure 4.4(a). The horizontal displacement of the monopile head has undergone three stages with the increase of horizontal load, i.e., (1) elastic deformation stage, (2) plastic deformation stage, (3) plastic failure stage. It was perceived from Figure 4.4(a) that the horizontal load and displacement were relatively small ($H < 1.9$MN, $h < 0.1$m) at the first stage, and the horizontal load and displacement curve of the monopile foundation presented a linear relationship. Meanwhile, the monopile-soil system was basically in a steady state. Once the horizontal displacement reached 0.1m, the horizontal displacement increased nonlinearly with the horizontal load until displacement reached 0.8m. Thus monopile-soil system entered the plastic deformation stage, and the stiffness of the pile head was only 13.5% of the initial stiffness. Subsequently, the relationship curve of horizontal load and displacement of the monopile foundation became linear again, indicating the seabed was in the plastic failure stage. The horizontal load-displacement curve of the monopile head varied gradually, and there were no distinct inflection points in the curve. Therefore, the monopile foundation in soft clay showed a progressive failure under unidirectional horizontal load, similar to the previous studies (He et al., 2019; Lombardi et al., 2015; He, 2016; Hong et al., 2017).

The distribution of equivalent-plastic strain of soil around the monopile foundation under different horizontal loads was presented in Figure 4.4(b). The plastic strain of seabed around the monopile foundation first appeared on the seabed surface in the horizontal load-imposed direction. The plastic failure zone gradually extended from the topsoil to deep soil with the horizontal load increasing until it reached the bottom of monopile foundation. The failure mode of the monopile foundation in soft-soil seabed under the ultimate horizontal load was "two-zone failure", as illustrated in Figure 4.4(c). Under the ultimate horizontal load, there were two visible failure zones in the seabed around the large-diameter monopole foundation, i.e., the wedge-shape failure zone and the circular rotation failure zone. The wedge-shape failure zone is mainly distributed in the shallow seabed (less than 3D). At the seabed surface, the influence range of horizontal load in front of the pile was mainly within 15m (3D). While 20m away from the pile, the topsoil displacement was small with little effect by pile disturbance. Besides, there was an apparent settlement in the seabed surface behind the pile (reverse direction of the horizontal load imposed). The circular rotation failure zone appeared around the monopile foundation toe, where the deep soil rotating around the pile toe formed a circular shear failure zone.

According to the classification criterion of piles under lateral loads proposed by Poulos and Hull (1989), the monopile foundation in this study belongs to a rigid foundation. The monopile foundation almost has no material deformation apparently, due to the significant stiffness of rigid monopile. Therefore, if the monopile is subjected to the ultimate horizontal load, the pile body rotates rigidly around a point in the pile body, leading to the "two-zone failure" in the pile-soil system, as shown in Figure 4.4(c). Rigid pile material can meet the requirements of bearing capacity without failure, and the failure of the pile-soil system only occurs in the soil around the pile. With the increase of horizontal displacement of the pile body, the plastic zone begins to appear in the shallow soil. It gradually develops into the deep soil until the whole soil bearing failure occurring, leading to the pile-soil system failure. Hong and He (2017) found the semi-rigid monopile foundation under cyclic horizontal load also showed two failure zones through centrifuge tests and numerical simulations. As the length of the monopile expands, the rigid pile becomes a flexible pile. Its failure mode under unidirectional and cyclic horizontal loads changes into "three-zone failure" from "two-zone failure".

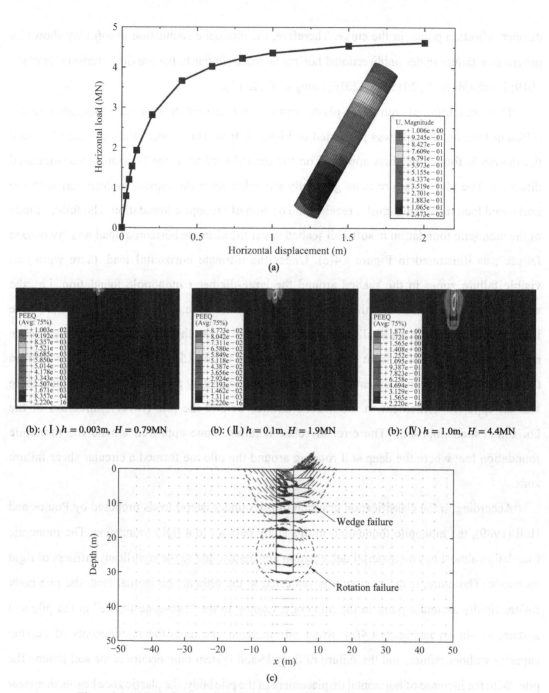

Figure 4.4 (a) The horizontal load-displacement response of monopile foundation,
(b) Distribution of equivalent-plastic strain of soil under different horizontal loads,
(c) The failure mode of monopile under the ultimate horizontal load (Dai et al., 2022)

4.1.2.2 Failure mode of monopile foundation under unidirectional vertical load

The monopile foundations of OWT are subjected to horizontal loads as well as vertical loads caused by the weights of superstructures, i.e., tower, rotor, nacelle, and blades. These vertical loads are the significant factors causing monopile settlement, so it is essential to study the bearing capacity and failure mode under vertical loads.

Different levels of vertical displacements were imposed on the pile head. The vertical load-displacement response of the monopile foundation was shown in Figure 4.5(a). It could be perceived that vertical displacement rose gradually with the increase of the vertical load. There was a distinct inflection point in the relationship curve of vertical load and displacement. When the vertical displacement was less than 0.1m, the relationship between load and settlement was linear approximately, while the relationship between load and settlement was nonlinear once settlement exceeded 0.1m. The distributions of the equivalent-plastic zone of soil under different vertical loads were illustrated in Figure 4.5(b). When the vertical load and displacement were small, the penetration of pile drove soil settlement around the pile, resulting in shear failure zone in the soil around the pile. With the increase of vertical displacement, the zone of plastic failure gradually extended from shallow soil into depth until reaching the pile end.

Figure 4.5(c) displayed the failure mode of monopile under the ultimate vertical load. Under the ultimate vertical load, a prominent shear failure zone was formed around the pile, where the soil particles mainly move downward. Because of the pile's penetration, the soil under the monopile foundation was compressed to form compression failure, resulting in vertical bearing capacity failure.

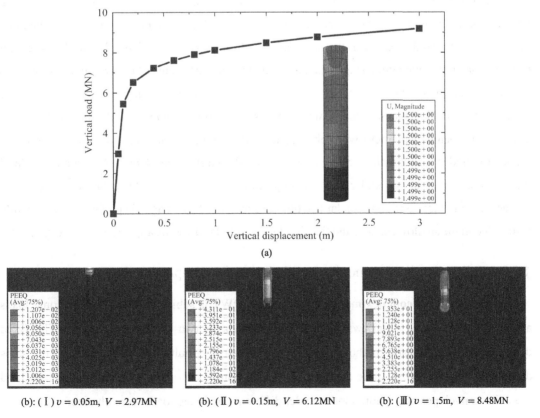

(a)

(b): (Ⅰ) $v = 0.05$m, $V = 2.97$MN   (b): (Ⅱ) $v = 0.15$m, $V = 6.12$MN   (b): (Ⅲ) $v = 1.5$m, $V = 8.48$MN

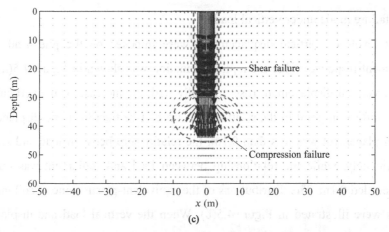

Figure 4.5 (a) The vertical load-displacement response of monopile foundation,
(b) Distribution of equivalent-plastic failure zone under different vertical loads,
(c) The failure mode of monopile foundation under ultimate vertical load (Dai et al., 2022)

4.1.2.3 Failure mode of monopile foundation under bending moment

The blades and tower structures of OWT were subjected to the long-term wind and wave loads. Those loads can be turned into bending moment loads on the monopile foundation. As a towering structure, OWT is prone to structural damage caused by the bending moment, affecting wind turbines' regular operation.

A series of different rotation angles were applied on a rigid reference point of the pile head, and the bending moment-rotation angle response was plotted in Figure 4.6(a). The bending moment-rotation angle response, the same as the aforementioned horizontal load-displacement response, has undergone three stages, i.e., a linear stage in early stage, then a nonlinear stage, and then turn back linear stage at last. The monopile foundation in soft clay showed a progressive failure under bending moment loads, which was similar to the response under horizontal loads. The ultimate bending moment was achieved as 100MN·m. The failure mode of the monopile foundation under the ultimate bending moment was shown in Figure 4.6(b). The failure mode of the monopile foundation under the bending moment was also a "two-failure zone", i.e., wedge-shape failure zone and circular rotation failure zone. The wedge-shape failure zone mainly distributed in the shallow seabed, and the circular rotation failure zone appeared in the deep seabed.

4.1.3 Failure mode of monopile foundation under complex loads

In the long-term complex marine environment, OWT is subjected to multi-directional loads and coupled action of wind, wave, and current loads in two or three directions. The fixed displacement loading ratio method was used to simulate the multidirectional loads. As the steps described in Section 4.1.1.3, the combined loads were imposed on the monopile foundation, and the failure envelopes and failure modes under different combined loads were obtained.

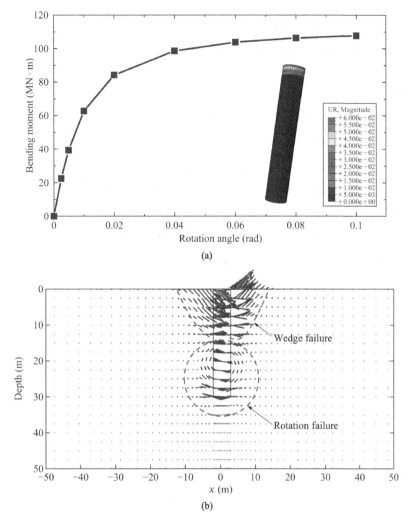

Figure 4.6 (a) Bending moment-rotation angle response of monopile foundation, (b) The failure mode of the monopile foundation under the ultimate bending moment (Dai et al., 2022)

4.1.3.1 Failure mode of monopile under combining loads of H-V and M-V

The displacement ratios of $\delta h/\delta v$ = 0.1, 0.25, 0.5, 1, 2, 4, 10 were imposed on the pile head separately. The failure points of all fixed displacement ratios were connected, forming the monopile foundation failure envelope. The normalized failure envelope in vertical and horizontal load space was shown in Figure 4.7(a). The bearing capacity in horizontal direction gradually increased slightly with the vertical load loading firstly, then reduced until reaching the ultimate vertical load. Under the action of H-V loads, according to the location of actual loads with respect to the failure envelope in the H-V load space, it could be estimated if the monopile foundation is in a failure state. If the actual point of vertical and horizontal coupling loads was located inside the failure envelope in the H-V load space, the service state of the monopile foundation was safe. However, the point was outside the curve, the monopile foundation was destabilized. If the actual point was precisely on the curve, monopile was in the critical state of failure. The fitting formula of failure envelope in the H-V load space was expressed as follows:

$$\left(\frac{V}{V_{\text{ult}}}\right)^5 + \left(\frac{H}{H_{\text{ult}}}\right)^2 = 1 \qquad (4.1)$$

where V was vertical load in kN; V_{ult} was the ultimate vertical load in kN; H was horizontal load in kN; H_{ult} was the ultimate horizontal load in kN.

The failure mode in the H-V load space was different from that under unidirectional vertical or horizontal load. Under the combined loads of vertical and horizontal load, one distinct soil flow mechanism could be identified for the large diameter monopile, namely a wedge failure near the monopile foundation. The wedge failure zone was different from that in the unidirectional horizontal and bending moment, expanding until monopile toe, as illustrated in Figure 4.7(b).

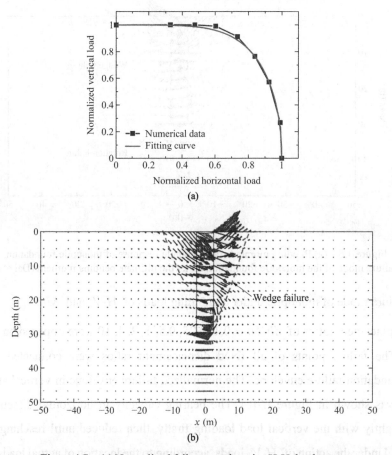

Figure 4.7 (a) Normalized failure envelope in H-V load space,
(b) The failure mode of the monopile foundation in H-V load space (Dai et al., 2022)

Like in the H-V load space, the failure envelope in M-V load space also was oval-shaped, as shown in Figure 4.8(a). Furthermore, the failure envelope in the M-V load space could be expressed by the Equation (4.2). The failure mode in M-V load space was also the wedge failure characteristic, illustrated in Figure 4.8(b).

$$\left(\frac{V}{V_{\text{ult}}}\right)^5 + \left(\frac{M}{M_{\text{ult}}}\right)^2 = 1 \qquad (4.2)$$

where V was vertical load in kN; V_{ult} was the ultimate vertical load in kN; M was bending moment in kN·m; M_{ult} was the ultimate bending moment in kN·m.

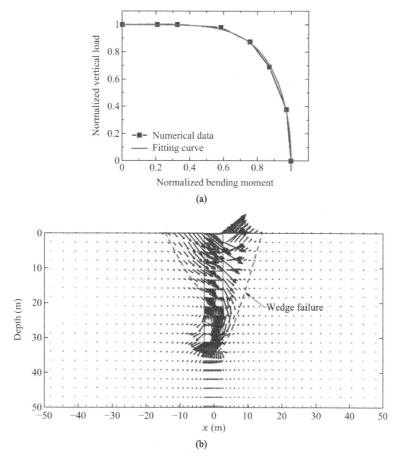

Figure 4.8 (a) Normalized failure envelope in M-V load space,
(b) Failure mode of the monopile foundation in V-M load space (Dai et al., 2022)

4.1.3.2 Failure mode of monopile under combining loads of H-M

The effect of horizontal load and bending moment on the monopile foundation was similar. Different displacement/rotation ratios were imposed on the pile head separately, and the failure points of all fixed displacement/rotation ratios were connected, forming the failure envelope. The failure envelope in M-H load space was linear, as shown in Figure 4.9(a), and the fitting formula of the failure envelope was expressed in Equation (4.3).

$$\left(\frac{H}{H_{\text{ult}}}\right) + \left(\frac{M}{M_{\text{ult}}}\right) = 1 \qquad (4.3)$$

where H was horizontal load in kN; H_{ult} was the ultimate horizontal load in kN; M was bending moment in kN·m; M_{ult} was the ultimate bending moment in kN·m.

The failure mode of the monopile foundation in M-H load spaces was the same as that in the unidirectional horizontal and bending moment. Two distinct soil flow mechanisms could be identified for the large diameter monopile, namely a wedge failure near the ground surface and rotational soil flow near the pile toe, as shown in Figure 4.9(b).

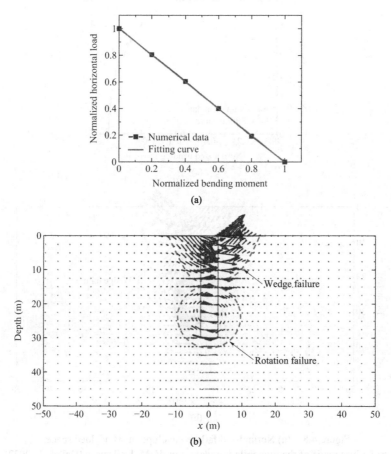

Figure 4.9 (a) Normalized failure envelope in M-H load space, (b) The failure mode of the monopile foundation in M-H load space (Dai et al., 2022)

4.1.3.3 Failure mode of monopile under combining loads of M-H-V

Some specific proportions of the ultimate vertical load were imposed on the pile head firstly, and the fixed displacement/rotation loading ratio method was then carried out in M-H load space. Then, the failure envelope under the M-H-V load space was obtained.

The specific proportions of ultimate vertical load (V_{ult}, $0.96V_{ult}$, $0.75V_{ult}$, $0.5V_{ult}$, $0.25V_{ult}$, $0V_{ult}$) were applied on the reference point on the pile head of the monopile firstly. The failure envelopes in M-H load space under different initial vertical loads were gained, as

Figure 4.10(a) shown. The variation tendency of failure envelopes under different initial vertical loads were similar. The areas of failure envelopes in *M-H* load space increased with the decrease of initial vertical load. When the initial vertical load was equal to the ultimate vertical load ($V = V_{ult}$), the monopile failure mode was the same as that under the unidirectional vertical load. Therefore, the failure envelope was located at the coordinate (0, 0, V_{ult}) in the *H-V-M* load space, and the failure envelope in the *H-M* load space degenerated into the origin point, as shown in Figure 4.11. Figure 4.10(b) presented the normalized failure envelope in three-dimensional *M-H-V* load space. The surface diagram of the failure envelope in three-dimensional could be expressed by the Equation (4.4).

$$\left(\frac{H}{H_{ult}}\right)^2 + \left(\frac{V}{V_{ult}}\right)^5 + \left(\frac{M}{M_{ult}}\right)^2 = 1 \tag{4.4}$$

The above expression of failure envelope in three-dimensional load space has the significance in assessing the service performance of a monopile foundation in the entire life cycle. Based on this method, the bearing capacity of monopile and stabilization of soft soil can be estimated simply by the actual location of real marine loads with respect to the failure surface. If the location of real horizontal, vertical, and bending moment loads exceeds the failure envelope surface, the monopile-soil system is in a failure state. If the point is just on the envelope surface exactly, it is on the critical state of failure. Whereas the point inside the failure envelop, it is in a stable state. Above method was universal for the determination of bearing capacity of monopile foundation, because the failure envelope was normalized by the ultimate bearing capacities under unidirectional loads. The change in seabed characteristics and pile diameters only brought the change of ultimate bearing capacities, not the normalized failure envelopes.

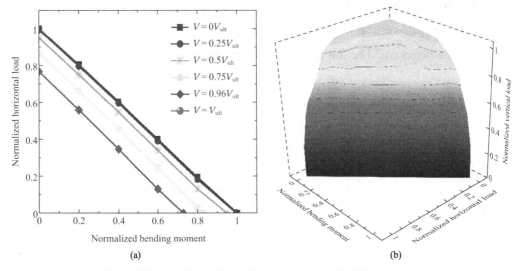

Figure 4.10 (a) Normalized failure envelope in *M-H-V* load space,
(b) Normalized failure envelope in three-dimensional *M-H-V* load space (Dai et al., 2022)

4.1.4 Effect of pile diameter on bearing capacity of monopile foundation

4.1.4.1 Effect of pile diameter on bearing capacity under unidirectional load

The diameter of the large-diameter monopile foundation for offshore wind generate engineering is typically 3-7m. The pile length $L = 30$m was kept constant, and the pile diameters were set as 3m, 4m, 5m, 6m, respectively. Furthermore, relationship curves of load and displacement with different diameters under different unidirectional loads were shown in Figure 4.11(a). The relationship curves of horizontal load and displacement of monopile foundation with different diameters were similar, all of which had grown slowly without a distinct inflection point. The monopile foundations with different pile diameters were progressive failure under different unidirectional horizontal load. With the enlargement of pile diameter (the buried depth remains at 30m), the horizontal bearing capacity increased. The diameter gradually increased from 3m to 4m, 5m, 6m, and the corresponding horizontal ultimate bearing capacity increased by 8.52%, 19.0%, 36.6%. The horizontal ultimate bearing capacity of monopile foundations with different diameters could be fitted by the following formula.

$$H_{ult} = 13.22\left(\frac{D}{L}\right) + 2.27 \tag{4.5}$$

where H_{ult} was ultimate horizontal bearing capacity, D was pile diameter, L_d was imbedded length.

The vertical load-displacement response and bending moment load-rotation angle response with different diameters were similar. As shown in Figure 4.11(b) and (c), with the increase of pile diameter (the buried depth remained at 30m), the vertical bearing capacity and bending moment increased. The ultimate vertical bearing capacity and bending moment of monopile foundations with different diameters could be fitted by the following formula.

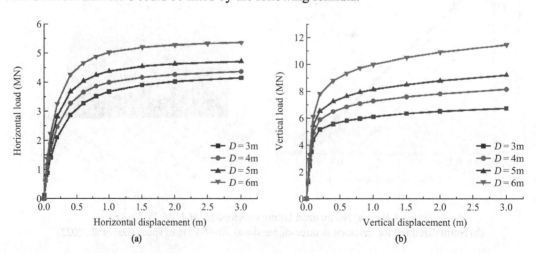

(a)   (b)

Chapter 4 Bearing characteristics of monopile foundations for offshore wind turbine

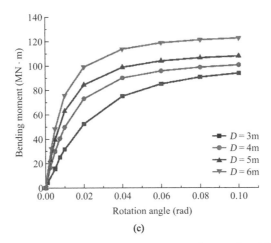

Figure 4.11 (a) The horizontal load-displacement response of monopile with different diameters,
(b) Vertical load-displacement response of monopile with different diameters,
(c) Bending moment-rotation angle response of monopile with different diameters (Dai et al., 2022)

$$V_{ult} = 40.12\left(\frac{D}{L}\right) + 2.20 \tag{4.6}$$

$$M_{ult} = 327.64\left(\frac{D}{L}\right) + 51.63 \tag{4.7}$$

where V_{ult} was the ultimate horizontal bearing capacity, M_{ult} was the ultimate bending moment.

4.1.4.2 Effect of pile diameter on bearing capacity under combined load

Based on the above soil-monopile foundation numerical modes with different diameters, the fixed displacement loading ratio method was used to investigate the bearing characteristics of the monopile foundation for OWT under complex loads. And the failure envelopes of monopile with different diameters under different loads were presented in Figure 4.12(a). The failure envelopes in H-V and M-V load space with different diameters were oval-shaped, and the failure envelopes in M-H load space were linear. With the enlargement of the monopile diameter, the area of failure envelope increased, and the bearing capacity increased. The normalization failure envelopes in H-V, M-V, and M-H load space could also be fitted using formula (Equation (4.1), (4.2), and (4.3)) in Section 4.1.

4.1.5 Summary

In this study, a 3D Finite Element model of the monopile foundation for OWT in soft soil was established. The bearing capacities and failure modes of monopile under unidirectional loads were explored firstly. Subsequently, the failure envelopes and failure modes in the load spaces of H-V, M-V, M-H, M-H-V were investigated. Finally, the effect of pile diameter on the bearing capacity under unidirectional and complex loads was discussed, respectively. The main research findings were summarized as follows.

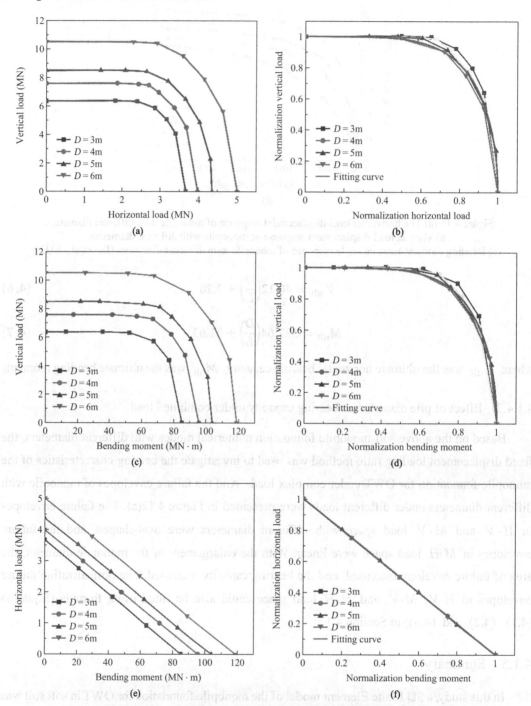

Figure 4.12 (a) Failure envelopes of monopile with different diameters in H-V load space,
(b) Normalization failure envelopes of monopile with different diameters in H-V load space,
(c) Failure envelopes of monopile with different diameters in M-V load space,
(d) Normalization failure envelopes of monopile with different diameters in M-V load space,
(e) Failure envelopes of monopile with different diameters in M-H load space,
(f) Normalization failure envelopes of monopile with different diameters in M-H load space (Dai et al., 2022)

(1) Two distinct soil flow mechanisms were identified for the large-diameter monopile, namely wedge failure near the ground surface and rotational soil flow near the pile toe under unidirectional horizontal load and bending moment.

(2) The failure envelopes of the monopile foundation were approximate "oval-shaped" in the H-V and M-V load space and "linear shaped" in the H-M load space.

(3) The failure envelope in the H-V-M load space provided a method to assess the service performance of the monopile foundation in its entire lifecycle and the stability of the seabed based on the actual location of real marine loads with respect to the failure surface.

(4) Enlarging pile diameter could enhance the bearing capacity of monopile foundations under unidirectional and multidirectional loads.

4.2 Dynamic response of offshore open-ended pile in sand under lateral cyclic loadings

4.2.1 Studies using large-scale model tests

4.2.1.1 Model test materials and methods

(1) Model box and soil sample preparations

Figure 4.13 shows the model box that is utilized for testing open-ended pile. The inner dimensions of the model box are 3m × 3m × 2m (length × width × height). The model box is provided with unloading port system as well as visual windows. Figure 4.14 shows the model particle gradation curve. Sand sample median grain size, non-uniformity coefficient and coefficient of curvature are 0.72mm, 4.25 and 1.47, respectively. The soil sample is controlled by sub-layered compaction method.

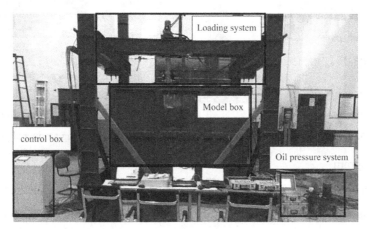

Figure 4.13　Model box

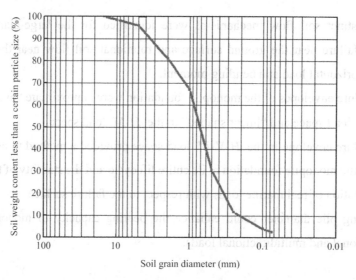

Figure 4.14 Mold particle gradation curve

(2) Model pile and sensor layout

The double-walled pile model consists of two concentric pipes of 6063 aluminum alloy material. The outer diameter, inner diameter, wall thickness and the pile length are 140mm, 120mm, 3mm and 1000mm respectively. Both the inner and outer tubes are instrumented with a fiber optic sensor (Figure 4.15(a)). The outer tube in addition is also instrumented with a soil pressure sensor (Figure 4.15(b)). Six installed soil pressure sensors are arranged in order from the pile bottom to the pile-top to measure the soil pressure in the pile-soil interface

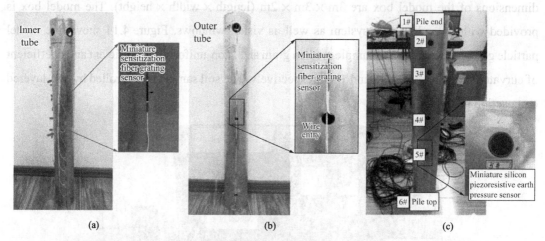

Figure 4.15 Instrumentation of model pile sensor layout for measuring strain and pressure

(3) Test program

Figure 4.16 shows that pile driving is proceeded using step loading method and the lateral cyclic loading is applied via a servo loading equipment. According to the proposed damage standard of laterally loaded pile (Cuellar, 2011), the ultimate bearing capacity of a single pile foundation under the lateral static load conditions can be the corresponding load when the pile top displacement reaches 0.1 times the pile diameter. Based on this calculation, the lateral ultimate bearing capacity of the pipe pile is 1587N. Leblanc et al (2010) defines two coefficients ζ_b and ζ_c to represent the characteristics of cyclic loading (Refer to Equations (4.8) and (4.9)).

$$\zeta_b = \frac{P_{max}}{P_R} \tag{4.8}$$

$$\zeta_c = \frac{P_{min}}{P_{max}} \tag{4.9}$$

Loads in ocean environment is usually not a completely regular load waveform, but for simplicity, the form of sine wave was adopted. The general loading form is shown in Figure 4.17. In this study, two types of cyclic loadings, i.e., $\zeta_c = -1$ and 0 are simulated. Table 4.2 summarizes the testing program for this study. It can be seen from the table that in total five experiments were conducted. These tests include 4 tests on open-ended pipe pile and one test on closed-ended pile. Among them, four tests were subjected to two-way loading while one test was subjected to one-way loading. The two-way load amplitudes adopted in this study are 200N, 500N and 800N, respectively, and the cyclic load ratios are 0.126, 0.315 and 0.504, respectively. The one-way cyclic load ratio is 0.113.

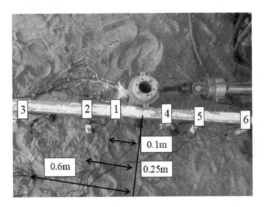

Figure 4.16 Cyclic loading control
(Note: 1-6 is the displacement meter number; the distance from the pile is 0.1m, 0.25m and 0.6m)

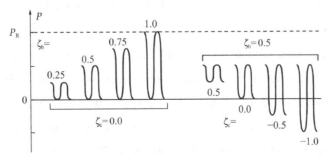

Figure 4.17 Schematic diagram of cyclic loading

Table 4.2 Test program

Test number	Pile diameter (mm)	Pile end	Loading method	Amplitude (N)
M1	140	open	two-way	200
M2	140	open	two-way	500
M3	140	open	two-way	800
M4	140	open	one-way	200
M5	140	closed	two-way	200

*Buried depth = 0.74m; Frequency = 4Hz; Cycles = 1000.

4.2.1.2 Model test results and discussion

(1) Measured pile top cumulative displacement under lateral cyclic loading

Figure 4.18 shows the variation in horizontal displacement with time for all different loading conditions (refer to open-ended pipe pile cases, M1-M4 in Table 4.2) is subjected to lateral cyclic loadings. It can be observed that trends of the displacement curves of the pile top are consistent under different cyclic loading modes. With the application of sinusoidal load, the displacement of the pile also changes sinusoidally over time.

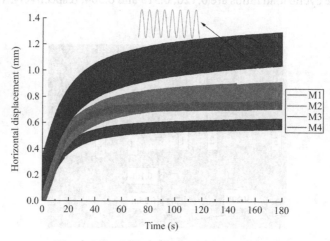

Figure 4.18   Pile top cumulative displacement

The total displacement for tests M1, M2, M3 and M4 are 0.65mm, 0.9mm, 1.35mm and 0.8mm, respectively. As observed from Figure 4.18, the maximum cumulative displacement of the pile top gradually increases with an increase in number of cycles and then gradually stabilizes beyond 100 cycles. The increase in horizontal displacement is much faster in first 100 cycles as compared to the cycles beyond. The displacement in first 100 cycles for M1, M2, M3 and M4 are about 76.9%, 74.4%, 81.4% and 78.9% of the total displacement, respectively. The cumulative displacement for M2 and M3 is higher than M1 by 27.7% and 51.8%, respectively. This is obviously due to an increase in load amplitude. It can be also observed that the cumulative displacement in case of one-way loading is 18.75% higher than two-way loading due to asymmetry of one-way loading.

In this study, the cumulative displacement under the lateral cyclic loading is predicted mainly by establishing the relationship (Equation (4.10)) between the displacement of the pile and the number of cycles. Hettler et al. (1981) carried out cyclic triaxial test and model pile test in dry sand. It is considered that the relationship between the lateral displacement (y_N) of the pile under the cyclic load, the displacement y_1 of the pile after the first cycle and the number of cycles N are as follows:

$$y_N = y_1(1 + C_N + \ln N) \tag{4.10}$$

where C_N is the weakening coefficient. For cohesion-less soil, C_N is usually 0.2. The weakening coefficients for M1, M2, M3 and M4 are found to be 0.159, 0.173, 0.181, and 0.186, respectively. The weakening coefficient is similar to that obtained by Zhu et al (2013).

(2) Measured load-displacement curve under lateral cycling load

Figures 4.19(a)-(d) shows the pile top load-displacement curves for cases M1-M4 under lateral cyclic loading. The ratio of the maximum load and the change in lateral displacement of the pile top is the lateral secant stiffness of the pile foundation. The stiffness of the soil around the pile changes under the cyclic load.

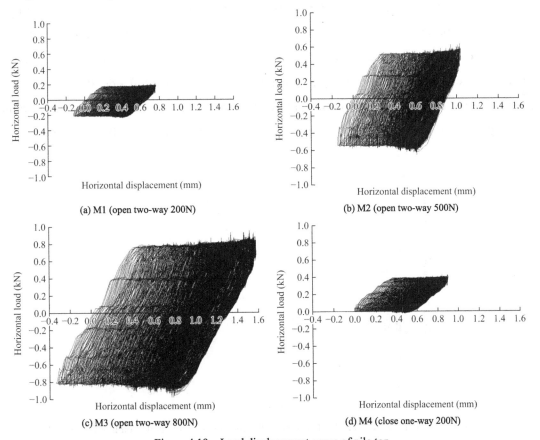

Figure 4.19 Load displacement curve of pile top

As seen from the figures, there is a hysteresis loop in load displacement curves of pile top with each cycle overlapping partially. Generally, for all the cases, the hysteresis loop is relatively small during first ten cycles of loading. The hysteresis loop gradually tilts toward the displacement axis. The area within the hysteresis loop curve is gradually reduced with an increase in cycles. Ultimately, the load displacement curve seems stabilized. The lateral stiffness for M1, M2, M3 and M4 decreases by 11.6%, 14.0%, 17.2% and 12.8% at the end of 1000 cycles. However, it should be noted that the major decrease in lateral stiffness for M1, M2, M3 and M4 are 6.97%, 9.9%, 13.5% and 8.01%, respectively in the first 100 cycles. These account for more than 60% of the overall decrease. It shows that cyclic loading can reduce the lateral secant stiffness of pile foundation. This is similar to the law obtained by Zhang et al. (2011). As per their law, the soil around the pile will "plastically" deform and gradually accumulate with an increase in loading cycles. The gradual deformation of soil will cause the weakening of the pile-soil system.

(3) Measured surface displacement under lateral cyclic loading

Figure 4.20 shows the variation in surface displacements (at points 1-6; refer to experimental set up in Figure 4.16) for case M1 (refer to Table 4.2). It can be observed that there was a rapid increase in displacement in the initial stage of application of cyclic load. However, the rate of increase reduces after around 5 seconds. The displacement for point 2 (gauge No.2; refer to Figure 4.16) on the right side of the pile is the largest, while on the left side, displacement for gauge No.5 is the largest. It can be observed that under the cyclic load, the soil will settle down in the range of 0.1m near the pile. Between the range of 0.1-0.25m, soil uplift will occur on both sides of the pile.

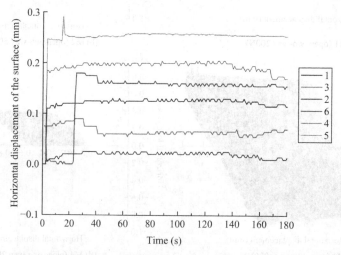

Figure 4.20 Variation of surface displacement with time under lateral cyclic loading for case M1 (open two-way 200N)

The displacement on the left side of the pile is larger than that on the right side, where the active pressure area is larger than the passive pressure area. The comparison shows that the maximum surface displacement of the pile under different loading conditions such as M1, M2, M3 and M4 are 0.25mm, 0.35mm, 0.4mm and 0.28mm, respectively. With an increase in cyclic load, the surface displacement increases gradually. Also, displacement under one-way cyclic load is larger than the two-way cyclic load. This law is similar to the cumulative displacement of the pile top (refer to Figure 4.19). The disturbance range of the soil around the pile is 2-3 times the diameter of the pile.

(4) Measured pile friction under lateral cyclic loading

Certain inclination occurs to pile under the application of lateral cyclic load to the pile. The lateral cyclic load makes changes in the direction of the pile, where lateral friction and lateral pressure of the pile differs with the traditional vertically loaded pile. The friction law of the pile body is similar under different loading conditions. This study provides mainly the curve of the frictional force of the pile body (for case M1 only) with the cycle period (as shown in Figure 4.21). As observed from the figure, the unit frictional force near the pile bottom decreases with an increase in number of cycles. The frictional force near the pile top tends to increase during the first 100 cycles. However, the change in frictional force is minimal in the middle of the pile body during the first 100 cycles. The friction at the pile bottom is generally weakened by about 3.8%, and the friction at the pile top is increased by about 3.4%. The friction of the pile body is generally found to reduce with the application of the lateral cyclic load; nevertheless, the decay rate is about 3.8%, where the degradation degree mainly accounts for more than 70% of the total degradation in the first 100 cycles.

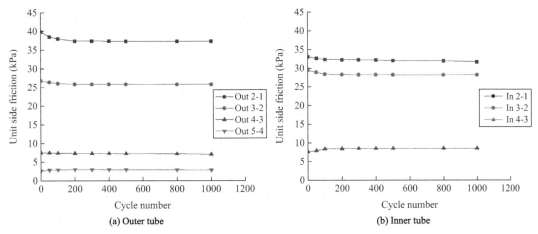

Figure 4.21 Unit side friction for case M1

The comparative analysis shows that the friction of the pile body is generally reduced with an application of the lateral cyclic load. The reduction tendency of friction is found to slow down with an increase in number of cycles. The overall reduction ranges of frictional forces for M1, M2, M3 and M4 are 3.6%, 3.8%, 4.2% and 3.7%, respectively. As the magnitude of the cyclic load increases, the frictional decay amplitude increases, and the frictional force in case of one-way cycle becomes weaker than two-way cycle. The friction inside the pile is mainly concentrated in the range of 2 times the pile diameter above the pile end, which can be called the "developing height" of the soil plug. In the range of "developing height" of the soil plug, the frictional force in the pile changes more obviously. The disturbance of the soil plug at the end of pile is directly proportional to the load amplitude. Moreover, the attenuation of friction on inner wall surface increases with load amplitude.

(5) Measured Lateral pressure of pile under lateral cycling load

Figures 4.22(a)-(d) illustrate the variation of soil pressure with depth for different loading conditions (M1-M4). From Figures 4.22(a), it is clear that the change in lateral pressure at a depth of 0.58m is zero. In general, for all cases, lateral soil pressure above the depth of 0.58m increases under cyclic load, while it decreases below the depth of 0.58m decreases. Hence, the depth of around 0.58m is the center of the pile rotation. The center of the pile rotation is located approximately at about 0.8 times of the pile depth. During the cyclic loading process, some soil pressure sensors on the pile body are positioned in the "active zone", and some of the sensors are located in the "passive zone". Definitions of "active zone" and "passive zone" are based on the first cycle; the load is applied to the right in the first half cycles, and the pile is tilted to the right. The left side of the pile is the passive zone, which is on the right side of the active zone.

Soil pressure sensors 4, 5 and 6 above the center of rotation are in the "passive zone", while the sensors 1 and 2 are in the "active zone". The lateral pressure of the active zone increases with the cyclic loading, while the lateral pressure of the passive zone decreases. The pressure in the passive zone increases with an increase in number of cycles. The major increase occurs during the first 100 cycles, accounting for more than 70% of the total. The lateral pressure of the active zone shows a decreasing trend. Analysis shows that the overall lateral pressure for M1, M2, M3 and M4 are attenuated by 6.9%, 7.5%, 8.8% and 7.3%, respectively.

(6) Measured static p-y curve under lateral cyclic loading

In this study, the sand p-y curve models proposed by American Petroleum Institute API (2000) and Reese et al. (1974) are used to calculate the horizontal load and displacement of monopiles. Figure 4.23 shows the variation of calculated load-displacement relationship in horizontal direction. As observed from Figure 4.23, the results obtained by two aforementioned methods are relatively closer to those of the static calculation (i.e., before the test cycle loading). With the application of cyclic load, the soil around the pile is disturbed and the ultimate bearing capacity of the soil after circulation is reduced. Ultimate bearing capacity of soil for cases M1, M2, M3 and M4 are reduced by about 11%, 14%, 17% and 13%, respectively. After 100 cycles, the results calculated by the two methods are quite different from the test results, indicating that the two static calculation methods cannot reflect accurately the influence of cyclic loading on displacement of piles.

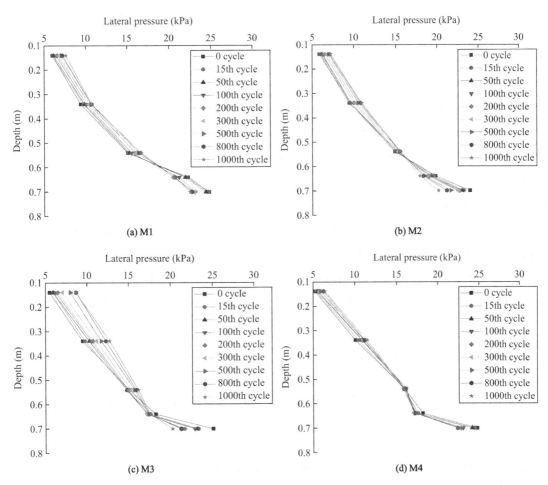

Figure 4.22   Variation of lateral soil pressures with depth for various loading conditions (M1, M2, M3 and M4)

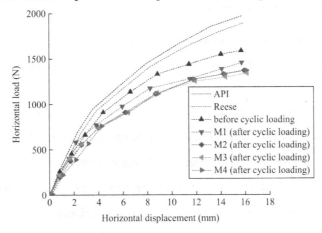

Figure 4.23   Static *p-y* curve based on *p-y* models proposed by API (2000) and Reese et al (1974)

4.2.1.3 Summary

In this study, the dynamic response of open-ended pipe piles under different loading modes in lateral cyclic loading is studied by large-scale indoor model tests. The main findings can be summarized as follows:

Both the increases of cumulative displacement on the pile top and the decrease of the lateral secant stiffness occur mainly in the first 100 cycles, which is in the range of 10%-25% but varies greatly with the change of the loading mode. One-way cyclic loading causes more lateral displacement than the two-way loading. Ultimate bearing capacity of the pile decrease logarithmically with the increase of the period, and the weakening coefficients are different for loading modes but all in the range of 0.15-0.2.

The cumulative displacement on the pile top increases with the increasing cyclic load ratio, but its increasing extent less than that of cyclic load ratio. Cumulative displacement reaches to around 1% of pile diameter when the cyclic load ratio increases to about 0.5.

The surrounding clear disturbance range of the soil is 2-3 times of the pile diameter, and the rotation center position of the pile body is about 0.8 times of the buried depth of pile body.

Both the soil plug and outer friction contributed significantly to pile lateral resistance, the "developing height" of the soil plug under lateral loading is in the range of 2 times the pile diameter above the pile end. The lateral pressure and frictional resistance of the active zone increases with the cyclic loading, while the lateral pressure and frictional resistance of the passive zone decreases with lateral loadings.

4.2.2 Numerical studies using Discrete Element simulation

4.2.2.1 Numerical model establishment

The DEM model used in this section is the same as the model described in Section 3.2.1.1 and the pile was generated using the same method as described in Section 3.2.1.2. Three open-ended rigid tubular piles were considered with the same wall thickness (2.475mm) and length ($L = 0.515$m) but different outer diameters (d_{pile} = 22.5mm, 45mm and 90mm for pile P1, P2 and P3, respectively). The three piles were first driven into the soil tank as described in Section 3.2.2.

It can be seen from Figure 4.24 that after the pile penetration is completed, a thin shear zone is formed next to the outer wall of the pipe pile. The shear zone mixes soil particles from different soil layers above. The soil particles in the shear zone have large horizontal and vertical displacements, resulting in large changes in the properties of the original soil. This indicates that the friction characteristics of the pile-soil interface not only depend on the local soil layer, but also on the upper soil layer. The width of the pile-soil interface shear zone is in the range of $2d_{50}$-$5d_{50}$ (d_{50}: median diameter). This is similar to the data obtained by Yang (2010) and White & Bolton (2004). The outer side of the shear zone is a severe disturbance zone of 2 times the pile diameter and a partial disturbance zone of 5 times the pile diameter. The soil particles in these two areas mainly undergo horizontal displacement. The outer part of the disturbed zone is a non-disturbed zone, and the soil around the pile is hardly affected. With the penetration of the pile, a part of the soil around the pile end is squeezed into the pile to form a soil plug, and a part of the particles are driven downward by the pile and distributed in the area adjacent to the side wall of the pile. During the process of pile penetration, the soil plug is not completely static, and the changes of soil layers of different colors can obviously reflect such phenomena. Since soil particles tend to move to a lower stress level, the ratio between the two parts of soil particles is related to the stress level around the pile. This phenomenon depends on the degree of soil blockage. The layering of the soil plug is consistent with the undisturbed soil layer, but its layered section is not flat. Most of the upper part presents the characteristics of an upward convex "active arch", while the soil at the bottom approximately equal to the pile diameter presents a concave "passive arch". It shows that "active arching" appears in most stages of pile penetration, which is mainly based on the constraint of the friction of the inner wall of the pile. The "passive arch" is the stage that is about to end. Corresponding to this, there is complete plugging at this time, and the active arch provides greater resistance. There is a thin shear zone between the soil plug and the inner wall, and the particle composition is not the same as that of the soil plug at the same level, but the confluence of particles in the upper soil layers.

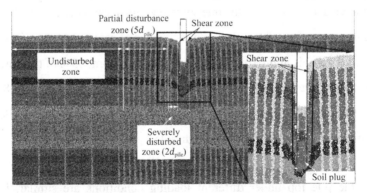

Figure 4.24 Installed pile and surrounding soil profile (Zhu et al., 2021)

4.2.2.2 Numerical simulation scheme

The buried depths of the model piles are around 0.4m, the model piles are located along the center line of the model box, the minimum distance between the model pile and the model box wall is greater than $7D$ (D is the pile diameter), and the distance between the pile end and the bottom of the box is greater than $4D$, and the boundary effect can be ignored. 100 cycles of lateral sinusoid cyclic load with a frequency of 40Hz was applied to the pile top to investigate the pile responses. Two-way symmetric cyclic load with amplitudes of 1000N, 3000N and 5000N were applied on the pile to explore the effect of load amplitude. One-way cyclic load with a maximum load of 1000N was also considered to show the effect of load pattern. The simulation program is shown in Table 4.3.

Numerical simulation parameters   Table 4.3

Test pile number	Pile diameter (mm)	Loading pattern	Amplitude (N)	Frequency (Hz)	Buried depth (m)	Loading cycle
P1	22.5	Two-way	1000	40	0.4	100
P2	45	Two-way	1000	40	0.4	100
P3	90	Two-way	1000	40	0.4	100
P4	45	Two-way	3000	40	0.4	100
P5	45	Two-way	5000	40	0.4	100
P6	45	One-way	1000	40	0.4	100

4.2.2.3 Analysis of the simulation results

(1) Pile displacement and rotation

Long-term lateral cyclic loading will cause the pile-soil force system to change, which will in turn result in rotation (tilt) of the pile. The upper limit for the maximum permanent rotation of a large diameter monopile of the offshore wind turbine at the mud line in Germany and the United Kingdom is 0.5° and 0.25°, respectively, while it is stipulated as 0.17° in China. Figure 4.25 shows the lateral cumulative displacement of the pile shaft at various depth after cyclic loading, indicating that the pile shaft rotates after 100 cycles, and the center of rotation (the position where the pile shaft displacement is zero) is approximately at the depth of 0.34m, 85% of the buried depth of pile shaft, which is similar to the research result of the center of rotation located about 85% of the burial depth of pile shaft proposed by Sun (2017). It is observed that the rotation of pile increases with decreasing pile diameter or increasing load magnitude.

Figure 4.26 shows the cumulative displacement of the pile top. Clearly, the cumulative displacement of the pile top under different loading conditions are similar. In all the cases, displacement first increases rapidly during initial 10 cycles and then gradually stabilizes. The main displacement occurred in the first 10 cycles accounts for about 71% of the total displacement, which bears a resemblance to the research results obtained by Chen (2021) that the first 10 cycles induced the majority of the response of the cumulative displacement of pile top. From P1 to P3, it can be seen that the larger the pile diameter, the smaller the cumulative displacement of the pile top. From P2, P4 and P5, it can be observed that the larger the load magnitude, the larger the cumulative displacement, as expected.

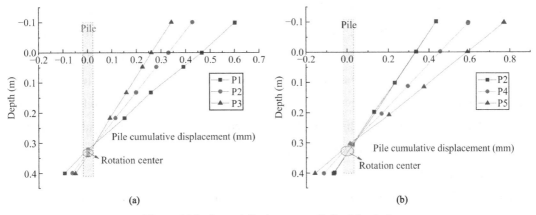

Figure 4.25 Lateral displacement of pile right shaft
(a) Various pile diameter; (b) Various load amplitude

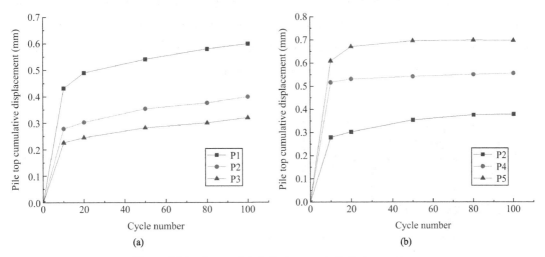

Figure 4.26 Accumulated displacement of pile top center
(a) Various pile diameter; (b) Various load amplitude

(2) Displacement of soil around the pile

Figure 4.27 shows the contour plot of the normalized displacement field in the entire soil container (Figure 4.27(a)) and the displacement vectors surrounding the pile P2 (Figure 4.27(b)). As is seen from the graphs, under cyclic loading, the soil around the piles is disturbed and the pile shaft tilts. From the contour plot of the displacement field, the influence zone of soil around the pile forms a "butterfly" shape, but the displacement in the "active zone" is larger than that in the "passive zone", where the "active zone" and "passive zone" are defined by the first quarter of the first loading cycle. Although the cyclic load set in the simulation test is symmetric, the pile rotates to the passive side of the first quarter of cyclic load, indicating that the permanent soil deformation in this first quarter of cyclic load dominates the direction of pile rotation. In the first quarter of the cycle, the load is applied rightwards, and the pile slants to the right, so the left side of the pile is active zone, and the right side is the passive zone. It is found through monitoring the simulation process that with the application of the first quarter of a sinusoidal load, pores are generated between the pile and soils behind the pile, and soil particles behind the pile move downwards to fill the pores, leading to the settlement of soils; on the other hand, soils in front of piles were pushed up, leading to soil heaves. In the next quarter, the load is reverted, but soils have less deformation leading to pile tilt to the right. After the next half of loading cycle, the residual soil deformation is settlement in the active zone and heave in the passive zone with pile tilt to the right, in spite of the symmetric sinusoidal loading applied. This type of soil deformation and pile rotation is accumulated in the following cycles as evident in the displacement vector diagram, Figure 4.27(b), as well as the pile shaft displacement as shown in Figure 4.25. It can also be observed that the soil plug has a slight displacement upward. With the increase of cyclic load (P4, P5), the upward movement of soil plug is slightly higher; while for one-way cyclic loading (P6), the upward movement of soil plug is slightly lower. It may be due to the large sloshing of the soil plug under the action of two-way cyclic loading, and the soil plug gradually becomes looser, resulting in a greater reduction in the friction between the soil plug and the inside of the pile.

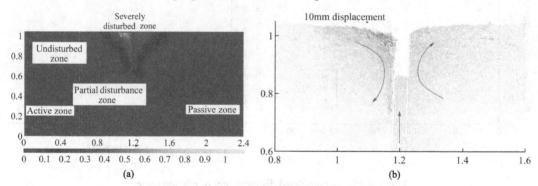

Figure 4.27 Displacement of soil around pile P2
(a) Normalized displacement field in the soil container; (b) Particle displacement near pile (Zhu et al., 2021)

(3) Load-displacement curve

Under the lateral cyclic loading, the interaction between pile and soil is weakening. Carrying normalization analysis through test and simulation results, the lateral load is divided by the corresponding load amplitude, and the lateral displacement is removed to maximum cumulative displacement. Figure 4.28 illustrates the load-displacement curves under normalized cyclic loading conditions. The curves basically resemble hysteresis loop. As the cycle period increases, the area of the loop gradually increases, indicating the increase in displacement of the pile top as well as reduction in lateral stiffness.

The lateral secant stiffness k_1 of the first cycle for P2, P4, P5 and P6 are 6.67kN/mm, 8.57kN/mm, 9.09kN/mm and 5.88kN/mm, respectively. The lateral secant stiffness k_{10} in the 10th cycle for P2, P4, P5 and P6 are 6.18kN/mm, 7.65kN/mm, 7.32kN/mm and 5.21kN/mm, respectively. The lateral tangential stiffness k_{100} in the 100th cycle for P2, P4, P5 and P6 are 5.88kN/mm, 7.32kN/mm, 6.94kN/mm and 4.84kN/mm, respectively. The overall reduction is 11.8%, 14.6%, 23.6% and 17.7%, mainly occurred in the first 10 cycles, accounting for 62.0%, 73.6%, 82.3% and 64.4% of the total.

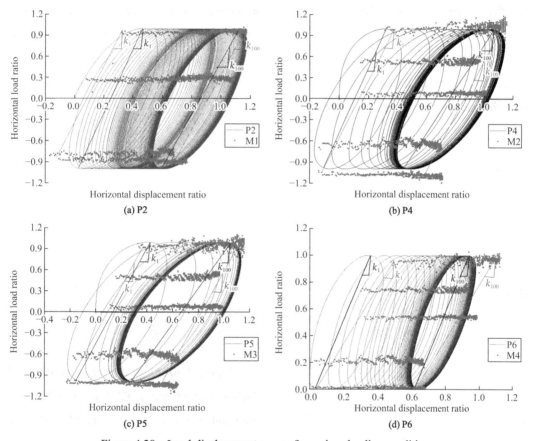

Figure 4.28 Load-displacement curves for various loading conditions

(4) Pile shaft frictional resistance

Figures 4.29(a) and (b) shows the variation of the pile frictional resistance with depth for outer and inner tubes, respectively for case P2. The side frictional resistance appears to increase along depth with some fluctuations for both outer and inner tubes of pile. It can be seen from Figure 4.29(a) that the side frictional resistance on the left side of the pile decreases at the vicinity of the pile top. The decreasing rate gradually becomes slower with the number of cycles. The side frictional resistance of the pile bottom tends to increase under application of load. The pile body rotates to the right around the center of rotation, and the friction between the soil and the pile on the left side of the pile top is reduced. The analysis of Figure 4.29(a) shows that the outer side frictional resistance above the center of rotation decreases. However, there is an increasing trend observed below the center of rotation. Variations in both these trends occur mainly during first 10 cycles. The outer side frictional resistance on the right side of the pile is opposite to the left side. As compared to Figure 4.29(a), the side frictional resistance (Figure 4.29(b)) on the left and right sides of the pile changes minimal during the period. Compared with the inner and outer frictional resistance, side frictional resistance of the right side is greater than side frictional resistance of the left side. The comparative analysis shows that the variation in lateral frictional resistance is highest in pile P5, while it is lowest for pile P2. For all the piles (P2, P4 and P5), the variations (i.e., gradually increase) of lateral frictional resistance occurs mainly during first 10 cycles of loading. Also, it can be stated that the lateral frictional resistance of the pile (based on P2 and P6) is more under the one-way cyclic load than the two-way cyclic load.

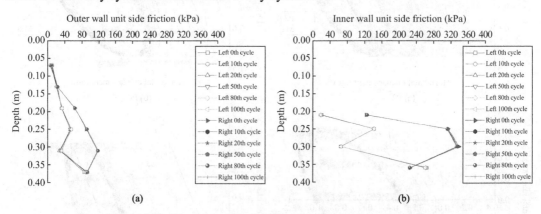

Figure 4.29   Evolution of shaft frictional resistance for various piles
(a) Outer wall; (b) Inner wall

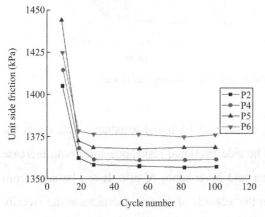

Figure 4.30   Unit side frictional resistance of pile body

Figure 4.30 shows the variation of the side frictional resistance of P2, P4, P5 and P6 with the number of cycles. As the number of cycles increase, the total side frictional resistance of the pile body decreases. The overall decrease for P2, P4, P5 and P6 are 3.4%, 3.8%, 5.1% and 3.5%, respectively. Generally, for all the cases (P2, P4, P5 and P6), the main decline inside frictional resistance in the first 10 cycles accounted for more than 77% of the total.

(5) Pile lateral pressure

With the application of sinusoidal cyclic load, the lateral pressure of the pile also changes in the form of sinusoidal wave. By dividing the inner and outer walls of the pile into small sections, the lateral force applied on each small section is monitored. Figure 4.31 shows the distribution of the inner and outer force of pile P2, where positive force represents rightward and negative force represents leftward. It can be seen that the distribution of lateral force exerted on pile foundation with the depth does not change remarkably under cyclic loading. In spite of the fluctuation of the distribution curves due to the discrete nature of soils, the inner lateral force generally increases with the depth of pile shaft while the outer lateral pressure generally increases first and then decreases, similar to the distribution law of soil pressure along pile shaft of rigid pile found by Klinkvort (2013). The lateral pressure mainly occurs in the first 10 cycles, and then the change tends to be stable. The lateral pressure at the depth of 0.33m, namely, the center of rotation, will reach its maximum. The force distribution between soil particles is shown in Figure 4.32. The thicker and denser lines represent the greater the soil forces. It can be seen from the figure that the inter-soil force generally shows an increasing trend as the depth increases. Since the diameter of the particles composing the pile is small, the particles of the soil sample are larger, and the number of particles in contact with the pile is uncertain. Therefore, the contact force between the pile and soil will also have a sudden change. This is also the main reason why the lateral stress fluctuates along the depth.

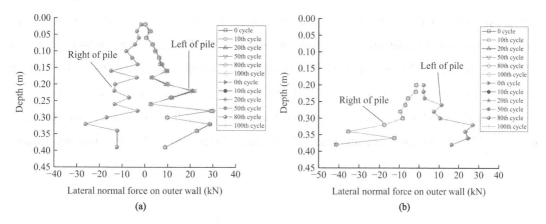

Figure 4.31 lateral force exerted on Pile P2

Figure 4.33 compares the distribution of total lateral pressure between piles P2, P4, P5 and P6 piles. It can be observed that the pressure of the pile-soil interface under one-way and two-way cyclic loading shows a reducing trend. An overall decrease of 8.9%, 9.3%, 10.1% and 9.7% was observed for piles P2, P4, P5 and P6, respectively. The major change in soil pressure occur mainly in the first 10 cycles, accounting for more than 74% of the total. Similar to the experimental results, the lateral pressure increases with the number of cycles. The major change occurs in the first 100 cycles accounting for more than 70% of the total. The lateral pressure of the active zone shows a decreasing trend. Comparative analysis showed that the overall lateral pressure for piles P2, P4, P5 and P6 was attenuated by 6.9%, 7.5%, 8.8% and 7.3%. It can be seen that the magnitude of the decline increases with an increase of the cyclic load ratio (i.e., lateral pressure higher in case of one-way loading than two-way loading).

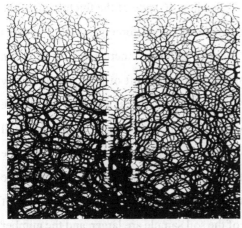

Figure 4.32 Soil particle contact force pattern (Zhu et al., 2021)

Figure 4.33 Comparison of computed unit lateral pressure distribution with cycles between P2, P4, P5 and P6

(6) Static p-y curve

Figure 4.34 shows the comparison of static calculation curves before and after cyclic loading between piles P2, P4, P5 and P6. It can be seen from Figure 4.34 that under the application of different cyclic loading modes, the lateral ultimate bearing capacity of P2, P4, P5 and P6 is reduced by 11.7%, 14.5%, 23.5% and 17.7%, respectively. The comparative analysis shows that the cyclic load can reduce the lateral bearing capacity of the pile foundation. The magnitude of the reduction under different loading conditions is different. The rate of decrease of amplitude is enhanced with the cyclic load ratio. The rate of decrease of amplitude is higher in case of on-way loading than two-way loading.

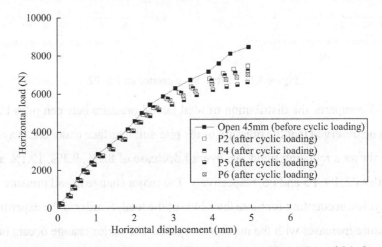

Figure 4.34 Comparison of simulated static load curves between P2, P4, P5 and P6 after circulation

Both measured and computed static load curves for all types of loading conditions were normalized by dividing the actual value (load or displacement) with the corresponding maximum value. Figure 4.35 shows the comparison of measured as well as computed normalized static load curves. It can be seen from the test results that with the application of cyclic load, the soil around the pile is disturbed, and the ultimate bearing capacity of the soil after circulation is reduced. The ultimate bearing capacity for M1, M2, M3 and M4 cycle is reduced by about 11%, 14%, 17% and 13%, respectively. In the simulation results, the lateral ultimate bearing capacity of the piles decreased by 11.7%, 14.5%, 23.5% and 17.7% after 100 cycles. The ultimate bearing capacity increases with the cyclic load ratio. The decreasing amplitude is gradually increased, and the reduction of bearing capacity in case of one-way cyclic loading is more than the reduction of two-way cyclic loading.

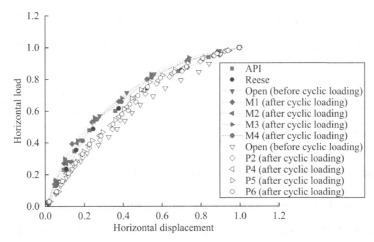

Figure 4.35 Comparison of measured and computed normalized static load curves between various loading conditions

4.2.2.4 Summary

In this section, the numerical simulation study on open-ended piles under lateral cyclic loading was carried out using Discrete Element modeling and the following conclusions have been drawn:

(1) The tilt of the pile shaft of the three simulated piles gradually accumulates to the right with the cyclic loading, with the center of rotation located at about 85% of the burial depth of the pile. The cumulative angle of the three simulated piles is less than 0.1°. With the application of the cyclic load, the disturbance of soil around pile produces plastic displacement, and the influence range of the soil around the pile presents a "butterfly shape", the displacement in the "active zone" is greater than that in the "passive zone".

(2) The horizontal stiffness of the open-ended pile gradually decreases with the cycle times, and the attenuation amplitude reaches more than 29%. The horizontal stiffness attenuation mainly occurs in the first 10 cycles, and the attenuation amplitude accounts for more than 80% of the overall attenuation. The reduction amplitude increases remarkably with one-way loading.

(3) The inner shaft frictional resistance supports the weight of the pile. Under cyclic loading, both the inner and outer shaft frictional resistance was weakened, with most attenuation occurred in the first 10 cycles. With the increase of the two-way cyclic load, the upward movement of soil plug is higher, while for one-way cyclic loading, the upward movement of soil plug is lower, possibly due to the less shaking of the soil plug.

(4) The lateral loading capacity of pile reduced following cyclic loading and this reduction is more obvious with higher load amplitude.

(5) For the static p-y curve, with the increase of load amplitude, the reduction of pile bearing capacity increases, and the reduction of one-way loading is greater than that of two-way loading.

4.3 Response of close-ended pile under lateral cyclic loading and the associated micro-mechanics

4.3.1 Summary of scaled model tests findings and field evidence

Extensive research has been carried out by Bhattacharya et al. (2013a, b), Lombardi et al. (2013) and Yu et al. (2015) to study the effects of cyclic loads on the first natural frequency of wind turbines. Tests were carried out on different types of foundations: monopiles, jackets, multiple pods. A typical test consists of the application of cyclic loading for a particular time interval (or for a certain number of cycles) and then measuring the frequency and damping of the system by a free vibration test. The cyclic loading was applied through an actuator. However, during the free vibration test (also known as a "snap back" test in the literature), the actuator was disconnected from the tower and the tower was given a small amplitude vibration and the acceleration of the system was recorded. The cyclic lateral loading was applied at different frequencies and for different lateral load magnitudes. This set of tests created a database of change of frequency and damping of the wind turbine system for different values of strain field in the soil next to the pile, forcing frequency imposed by the different dynamic loads, and number of cycles of loading.

Figure 4.36 shows one such graph obtained from scaled model test studies by Bhattacharya et al. (2013a) and Yu et al. (2015) on monopiles where the observed change in natural frequency is plotted with the number of cycles for various levels of strains in the soil next to the pile. The main observations from the tests are as follows.

- For strain-hardening sites (e.g. loose to medium dense sand) where the stiffness of the soil increases with cycles of loading, the natural frequency of the overall system will increase possibly due to densification.
- For strain-softening sites (clay sites) where the stiffness of the soil may decrease with cycles of loading, the natural frequency of the overall system will also decrease correspondingly. Of course, this depends on the strain level in the soil next to the pile and the number of cycles.

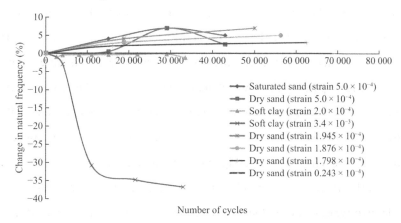

Figure 4.36 The observed change in natural frequency with the number of cycles for different strain levels in the soil around the pile (Cui and Bhattacharya, 2016)

There is some field evidence of dynamic soil-structure interaction and they are briefly discussed below.

- Natural frequency was measured for a wind turbine structure in the Hornsea wind farm where it was observed that the natural frequency decreased from 1.23 to 1.13Hz after 3 months of operation (Lowe, 2012).
- Kuhn (2000, 2002) reported that the target design frequency for Lely wind farm of 0.4 increased to 0.63Hz after 6 years of service.

4.3.2 Numerical findings using discrete element method modeling

4.3.2.1 Description of discrete element method model

Numerical simulations were performed by Cui and Bhattacharya (2016) to investigate the underlying mechanism for soil stiffness changes surrounding the wind turbine monopile. The discrete element method (DEM) was found to be more appropriate than other numerical methods (e.g. finite-element method) as it allows direct monitoring of change in soil stiffness, and more importantly it offers a method to analyze the micromechanics, which underlies the stiffness changes. In their study, the elastic Hertz-Mindlin contact model (Mindlin and Deresiewicz, 1953) is adopted. An opensource DEM code modified and validated in previous studies (Cui, 2006; Cui et al., 2007; O'Sullivan et al., 2008) was used to perform their study.

To simulate the soil stiffness changes, a DEM model of a soil tank (100mm × 100mm × 50mm) was first created by Cui and Bhattacharya (2016). The soil tank was filled by about 13000 spherical particles with radii in the range of 1.1-2.2mm. The particles were deposited under gravity. The pile is 20mm in diameter and was embedded to a depth of 40mm by removing particles located in the space which was to be occupied by the pile. Particles were allowed to settle down again following the installation of the pile. Their work aims to obtain qualitative characteristics of soil behaviors for investigations of micromechanics, not to reproduce quantitatively the model tests. Therefore, they used large particle sizes, which may cause size effect. Once the soil particles were settled down in the soil tank, cyclically horizontal movements were assigned to the pile to simulate the cyclic movements of OWT monopile due to the cyclic loadings. Translational movements rather than rotational movements were assigned to the pile at this stage. Three different strain amplitudes, 0.1, 0.01 and 0.001%, were chosen to examine the effects of strain levels. For each strain amplitude, two types of cyclic loading were applied: symmetric cyclic loading with stains in the range of (−0.1, 0.1%), (−0.01, 0.01%) and (−0.001, 0.001%) and asymmetric cyclic loading with stains in the range of (0, 0.2%), (0, 0.02%) and (0, 0.002%). As constrained by the computational costs, 500 cycles were simulated by Cui and Bhattacharya (2016) for strain amplitude of 0.1% and 1000 cycles were simulated for other strain amplitudes. The simulation for 0.1% strain amplitude required about 1 month in computational time. The simulation parameters used are listed in Table 4.4.

Input parameters for DEM simulation (Cui and Bhattacharya, 2016)   Table 4.4

Parameters	Value
Soil particle density ρ_s (kg/m³)	2650
Particle sizes (mm)	1.1, 1.376, 1.651, 1.926, 2.2
Inter-particle frictional coefficient μ	0.3
Particle-boundary frictional coefficient μ	0.1
G_s (Hertz-Mindlin contact model) (Pa)	2.868×10^7
Poisson's ratio	0.22
Initial void ratio e	0.539

4.3.2.2 Stress-strain response and damping ratio

The resultant horizontal stress applied on the pile against the horizontal strain of soil is illustrated in Figure 4.37. As shown in Figure 4.37, the stress-strain curves form hysteresis loops, indicating the energy dissipations during cyclic loading. It is also observed that the areas of the hysteresis loops increase with strain amplitude, indicating greater energy dissipation. The hysteresis damping ratio, α, can be determined by the expression (Karg, 2007)

$$\alpha = \frac{A}{4\pi A_\Delta} \tag{4.11}$$

where A is the area of the hysteresis loop, representing the energy dissipated and A_Δ is the area of the triangle as indicated in Figures 4.37(a) and 4.2(b), representing the elastic energy stored in the soil during one load cycle. The variation of the damping ratio during cyclic loading is illustrated in Figure 4.38. The damping ratio for the strain amplitude of 0.01% oscillated about a constant value. However, it is obvious that the damping ratio for the strain amplitude of 0.1% decreased dramatically in the first 30 cycles and then oscillated at an approximately constant value. The damping ratios for asymmetric cyclic loading are much lower than those for the corresponding symmetric cyclic loading, due to the higher elastic energy stored in each cycle.

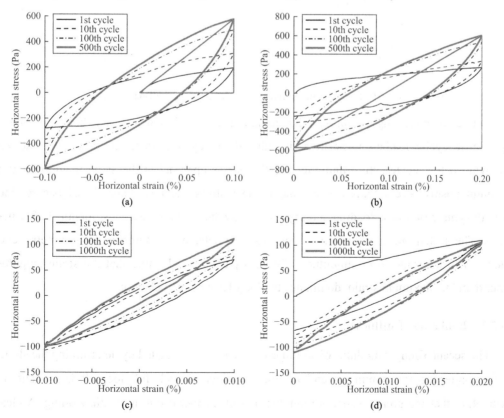

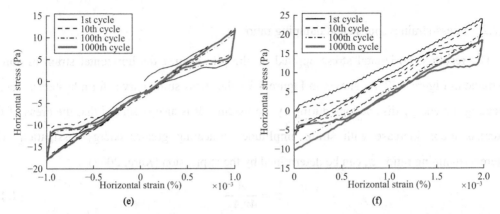

Figure 4.37   Hysteresis loops formed by the stress-strain curves during cyclic loadings
(a) Strain (−0.1, 0.1%); (b) Strain (0, 0.2%); (c) Strain (−0.01, 0.01%); (d) Strain (0, 0.02%);
(e) Strain (−0.001, 0.001%); (f) Strain (0, 0.002%) (Cui and Bhattacharya, 2016)

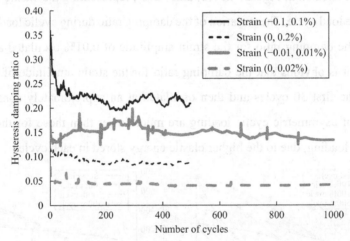

Figure 4.38   Hysteresis damping ratio in each cycle (Cui and Bhattacharya, 2016)

It is also interesting to observe that although the stresses in the first half cycle for the asymmetric cyclic loading is positive, it reduced to negative when the strain goes to zero. Following a few cycles, the minimum negative stress approaches the same magnitude as the maximum positive stress. Moreover, the magnitude of stresses and the shape of the hysteresis loops for both symmetric cyclic loading and asymmetric cyclic loading are almost identical after many cycles. The system under asymmetric cyclic loading behaves the same as the symmetric cyclic loading with the same strain amplitude after many cycles, indicating that the strain amplitude, rather than the maximum strain, dominates the long-term cyclic behavior.

4.3.2.3   Evolution of stiffness

The secant Young's modulus of soil in each cycle was calculated by determining the slope of a line connecting the maximum and minimum points of each full loop. It is evident from Figure 4.37 that the secant Young's modulus of soil increased during cyclic loading. A clearer evolution of secant Young's modulus is shown in Figure 4.39. At a strain amplitude of 0.1%, Young's modulus increases dramatically from 250kPa to around 600kPa for symmetric cyclic loading and from 130kPa to around 600kPa for asymmetric cyclic loading. At a strain amplitude of 0.01%, the Young's modulus increases quickly in the first few cycles and then only increases slightly to about 1100kPa. At a strain amplitude of 0.001%, Young's modulus only increases initially then mobilizes at a constant value at 1500kPa. The initial stiffness of asymmetric cyclic loading is lower than that of symmetric cyclic loading with the same strain amplitude due to a higher maximum strain applied. However, following a few cycles, the stiffness for both types of cyclic loading approaches the same values, confirming the observations from the stress-strain responses.

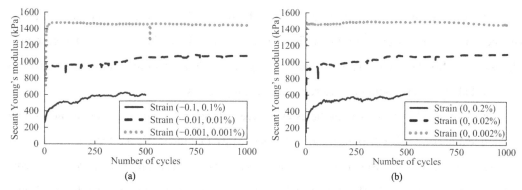

Figure 4.39 Secant Young's modulus of soil at the end of each cycle
(a) Symmetric cyclic loading; (b) Asymmetric cyclic loading (Cui and Bhattacharya, 2016)

The Young's modulus against horizontal strain in the first half cycle and in the 500th cycle for a strain amplitude of 0.1% are shown in Figure 4.40. The stiffness-strain curve in the first cycle displays the similar "S" shape as expected for the shear modulus-shear strain curve. Following cyclic loadings, stiffness at different strain levels all increases significantly. At a strain amplitude of 0.01 and 0.001%, stiffness also increases but at a much smaller scale as indicated in Figure 4.39.

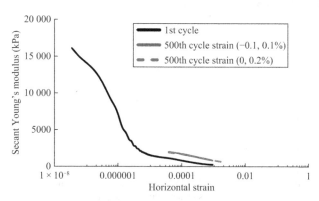

Figure 4.40 Secant Young's modulus against horizontal strain (Cui and Bhattacharya, 2016)

4.3.2.4 Convective granular flow

In the model tests (e.g. Bhattacharya et al., 2013a), ground settlements around the monopile could be observed. To illustrate the ground settlement, plots of incremental soil particle displacements in the symmetric cyclic loading with a strain amplitude of 0.1% in the first 50 cycles and in the next 50 cycles are examined by Cui and Bhattacharya (2016) as given in Figure 4.41 The soil particle displacements in the symmetric cyclic loading with a strain amplitude of 0.01% were also illustrated for comparison. Each arrow in the plot starts from the original center of a particle and ends at the new center at the end of a given cycle. It is evident that the soil particles surrounding the pile moved downwards, causing ground settlement. Soil densification around the pile is the main reason causing the increase of soil stiffness. It is also clear that the soil particle displacements are only significant in the first 50 cycles, underlying the significant increase in soil stiffness in the first 50 cycles. The soil particle displacements for a strain amplitude of 0.01% are less remarkable, corresponding to less reduction in soil stiffness.

It is also evident that only soil particles in an inverted triangular region enclosing the monopile have noticeable convective displacements. This is because that the vertical effective stress in a soil body increases linearly against depth, which imposes an increasing constraint on particle movement against depth. Cuéllar et al. (2012) performed physical tests of cyclic rotation of a pile and observed the granular convective flow and soil densification around the pile. The shape of the convective soil volume observed by them is similar to the region where the soil particle displacements were concentrated in the DEM simulations performed by Cui and Bhattacharya (2016). In the DEM simulations, samples with same initial void ratio of 0.539 (medium dense sample) all showed densification behavior and stiffness increase under cyclic loading. Cui and Bhattacharya (2016) also stated that it would be interesting to investigate soil behaviors and stiffness evolutions for a wide range of initial void ratios.

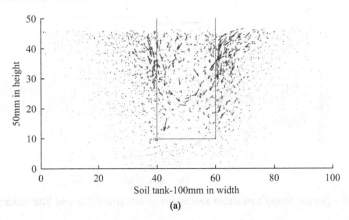

(a)

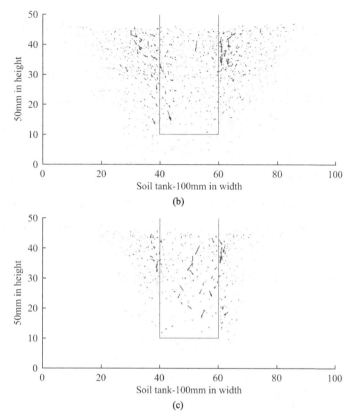

Figure 4.41 Incremental soil displacements at the end of the given cycle (unit: mm)
(a) Strain (−0.1, 0.1%), 1st-50th cycles; (b) Strain (−0.1, 0.1%), 51st-100th cycles;
(c) Strain (−0.01, 0.01%), 1st-1000th cycles (Cui and Bhattacharya, 2016)

4.3.2.5 Contact stresses and forces

Average radial stresses

The evolution of average radial stress on the pile at the end of each cycle is depicted in Figure 4.42 Due to soil densification, the radial stress increased significantly for a strain amplitude of 0.1% under cyclic loading, reached a peak value and then mobilized at about a constant value. Increased radial stress would increase side friction and would improve shaft resistance. The increase in radial stress is less remarkable for smaller strain amplitudes due to smaller particle displacements.

The evolutions of average radial stress in representative cycles are illustrated in Figure 4.43 The shapes of the radial stress-strain curves are quite different for different strain amplitudes. For symmetric cyclic loading with 0.01% strain amplitude (Figure 4.43(c)), the radial stress increases at positive strain values, but decreases slightly at negative strain values. However, for 0.1% strain amplitude (Figure 4.43(a)), the radial stress increases at both positive and negative strains, forming a "butterfly" shaped curve. It is also interesting to observe that, for asymmetric cyclic loading

(Figure 4.43(b)), the stress-strain response in the first few cycles is different from that in the correlated symmetric cyclic loading, However, it eventually evolves to the same "butterfly" shape after many cycles. It confirms again that the influence of different maximum strains of cyclic loading can be eliminated after many cycles and the dominant factor for cyclic soil response is the strain amplitude. A comparison between the patterns of convective particle flow was made; however, no noticeable difference can be identified between the particle flow patterns. In other words, particle flow is not the main reason contributing to the change in the shape of the radial stress-strain curves for symmetric cyclic loading. Further study will seek the underlying mechanism from the perspective of microstructure/fabric of particles. Asymmetric response of radial stresses may suggest unbalanced radial stresses at both sides of the pile. Further investigation on the unbalanced horizontal forces is provided as follows.

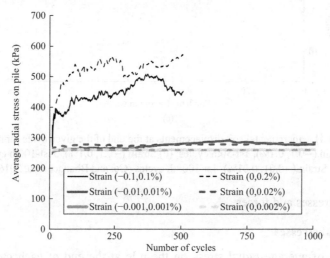

Figure 4.42 Evolution of average radial stress on the pile at the end of each cycle (Cui and Bhattacharya, 2016)

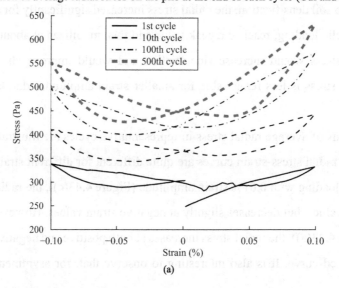

(a)

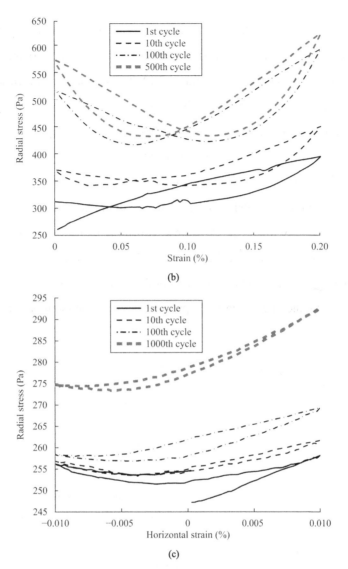

Figure 4.43　Evolution of average radial stresses in representative cycles
(a) Strain (−0.1, 0.1%); (b) Strain (0, 0.2%); (c) Strain (−0.01, 0.01%) (Cui and Bhattacharya, 2016)

Unbalanced horizontal force

As shown in the hysteresis loops formed by the cyclic stress-strain curves in Figure 4.37, at the end of each cycle, the stress does not return to zero but builds up gradually. The evolution of unbalanced horizontal force at the end of each cycle is illustrated in Figure 4.44. It is evident that the unbalanced horizontal force is more significant with increasing strain amplitude. It can also be seen that the unbalanced force for asymmetric cyclic loading is much larger in magnitude and is oriented in the opposite direction compared with that for symmetric cyclic loading. It is because that for asymmetric cyclic loading, the pile only moves to the right side and compresses the soils on the right side significantly; therefore, the horizontal force on the pile at the end of each cycle is oriented to the left. However, for the symmetric cyclic loading, the pile compressed the soils on both sides to the same strain level. At the end of a full cycle, the residual horizontal force is oriented to the right. By referring to Figures 4.37(a) and 4.37(b), it can be understood clearly that the end of a symmetric loading cycle is in the middle of a hysteresis loop, but the end of an asymmetric loading cycle is at the left corner of the hysteresis loop, even though the stress magnitudes for both hysteresis loops are approximately the same. Therefore, although the asymmetric cyclic loading does not cause different dynamic soil responses (stress-strain curves) from the symmetric cyclic loading during the cyclic loading, it does build up greater unbalanced horizontal force, which will cause excessive tilt and undermine the stability of the monopile more significantly in the long term. In the study by Cui and Bhattacharya (2016), monopile was driven to move by a predefined constant velocity. It is more realistic to simulate the free motions of monopile under the action of resultant external force/moment, including the interaction force/pressure between the monopile and the soil.

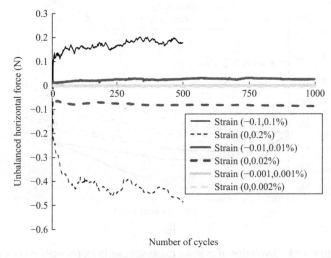

Figure 4.44 Evolution of unbalanced horizontal force on the pile at the end of each cycle
(Cui and Bhattacharya, 2016)

4.3.3 Summary

The DEM simulations and small-scale tests provide a good understanding on the soil-structure interaction of foundations of OWTs. Various features observed in model tests has been replicated by Cui and Bhattacharya (2016) in DEM studies and thus provides confidence in small-scale physical model tests. The following conclusions could be drawn from their study.

- Stiffness of granular soils increases under cyclic loading. Therefore, the stiffness of monopiles founded on granular material is expected to increase with cycles of loading and this increase may cause a change in natural frequency of the wind turbine system. It may be concluded that a "soft-stiff" will move towards the 3P frequency. It is necessary for the designers to predict the change in frequency which is essential to predict the fatigue life.

- The convective soil flow and the soil densification surrounding the monopile is the main reason underlying the increase in stiffness. These phenomena are more significant with the increase in strain amplitude.

- Due to soil densification, the average radial stress on the pile, and thus the shaft resistance of the pile increases under cyclic loading.

- Asymmetric cyclic loading applies larger maximum strain to soil, which results in higher stress in the first cycle; however, the difference in the magnitude of stresses and the shape of the stress-strain curves between asymmetric cyclic loading and symmetric cyclic loading eliminates quickly under cyclic loading. The long-term governing factor for the dynamic stress-strain response is the cyclic strain amplitude.

- The asymmetric cyclic loading builds up greater unbalanced horizontal force, which will undermine the stability of the monopile more significantly in the long term.

References

Abdel-Maksoud M. Substantial research secures the blue future for our blue plant. J. Renew. Energy Sustain. Dev. (RESD), 2016, 2: 4-5.

Abhinav K A, Saha N. Coupled hydrodynamic and geotechnical analysis of jacket offshore wind turbine. Soil Dynamics and Earthquake Engineering, 2015, 73: 66-79.

American Petroleum Institute. Recommended practice for planning, designing and constructing fixed offshore platforms-working stress design. 21st ed. Washington D C: American Petroleum Institute, 2000.

Anagnostopoulos C, Grorgiadis M. Interaction of axial and lateral pile responses. Journal of Geotechnical Engineering, 1993, 119: 793-798.

Bhattacharya S, Cox J A, Lombardi D, et al. Dynamics of offshore wind turbines on two types of foundations. Proceedings of the Institution of Civil Engineers: Geotechnical Engineering, 2013a, 166(GE2): 159-169.

Bhattacharya S, Nikitas N, Garnsey J, et al. Observed dynamic soil-structure interaction in scale testing of offshore wind turbine foundations. Soil Dynamics and Earthquake Engineering, 2013b, 54: 47-60.

Bolton M D, Cheng Y P. Micro-geomechanics. Constitutive and centrifuge modelling: Two extremes (ed. SM Springman), 2001: 59-74.

Broms B B. Lateral resistance of piles in cohesive soils. Journal of the Soil Mechanics and Foundations Division, 1964, 90: 27-64.

Chen R P, Ren Y, Chen Y M. Experimental investigation on a single stiff pile subjected to long-term axial cyclic loading. Chin. J. Geotech. Eng., 2011, 33(12): 1926-1933. (In Chinese)

Cuéllar P, Georgi S, Baeßler M, et al. On the quasi-static granular convective flow and sand densification around pile foundations under cyclic lateral loading. Granular Matter, 2012, 14(1): 11-25.

Cuéllar P. Pile foundations for offshore wind turbines: Numerical and experimental investigations on the behaviour under short-term and long-term cyclic loading. University of Technology Berlin, 2011.

Cui L, Bhattacharya S. Soil-monopile interactions for offshore wind turbines. Proceedings of the ICE-Engineering and Computational Mechanics, 2016, 169(4): 171-182.

Cui L, O'Sullivan C, O'Neil S. An analysis of the triaxial apparatus using a mixed boundary three-dimensional discrete element model. Geotechnique, 2007, 57(10): 831-844.

Cui L. Developing a virtual test environment for granular materials using discrete element modelling. PhD. Thesis, University College Dublin, Ireland, 2006.

Dai S, Han B, Huang G X, et al. Failure mode of monopile foundation for offshore wind turbine in soft clay under complex loads, Marine Georesources & Geotechnology, 2022, 40: 1, 14-25.

Dai S, Han B, Liu S L, et al. Neural network-based prediction methods for height of water-flowing fractured zone caused by underground coal mining. Arab J Geosci, 2020, 13: 495.

Duan N, Cheng Y. A modified method of generating specimens for a 2D DEM centrifuge model. Geo-Chicago, American Society of Civil Engineers, 2016: 610-620.

Duan N. Mechanical characteristics of monopile foundation in sand for offshore wind turbine. UCL: University College London, 2016.

Esteban M, Leary D. Current developments and future prospects of offshore wind and ocean energy. Appl Energy, 2012, 90(1): 128-136.

EWEA. The economics of wind energy. 2009.

Fan H Y. Dynamics and fatigue analysis on monopile foundation of offshore wind turbine under wind and wave loads. Harbin Institute of Technology, 2016.

Fan K F, Li W P, Wang Q Q, et al. Formation mechanism and prediction method of water inrush from separated layers within coal seam mining: A case study in the Shilawusu mining area, China. Engineering Failure Analysis, 2019, 103: 158-172.

Georgiadis M, Anagnostopoulos C, Saflekou S. Centrifugal testing of laterally loaded piles in sand. Canadian Geotechnical Journal, 1992, 29(2): 208-216.

Guo W D, Ghee E H. Behavior of axially loaded pile groups subjected to lateral soil movement. Foundation Analysis and Design: Innovative Methods, 2006: 174-181.

GWEC. Global Wind Report Annual Market Update. 2018.

Han B, Zdravković L, Kontoe S. The stability investigation of the Generalised-α time integration method for dynamic coupled consolidation analysis. Computers and Geotechnics, 2015, 64: 83-95.

Hazzar L, Hussien M N, Karray M. On the behaviour of pile groups under combined lateral and vertical loading. Ocean. Engineering, 2017, 131: 174-185.

He B, Lai Y Q, Wang L Z, et al. Scour Effects on the Lateral Behavior of a Large-Diameter Monopile in Soft Clay: Role of Stress History. Journal of Marine Science and Engineering, 2019, 7: 170.

He B. Lateral behavior of single pile and composite pile in soft clay. Zhejiang University, Hangzhou, 2016.

Hettler A. Verschiebung Starrer und Elastischer Gründungskörper in Sand bei Monotoner und Zyklischer Belastung. Ver"oentlichungen des Institutes für Bodenmechanik und Felsmechanik der Universität Fridericiana in Karlsruhe, Deutsch-land, Heft 90, 1981.

Hong Y, He B, Wang L. Cyclic lateral response and failure mechanisms of a semi-rigid pile in soft clay: centrifuge tests and numerical modelling. Canadian Geotechnical Journal, 2016: 0356.

Karg C. Modelling of Strain Accumulation Due to Low Level Vibrations in Granular Soils. PhD thesis, Ghent University, Ghent, Belgium, 2007.

Karthigeyan S, Ramakrishna V V G S T, Rajagopal K. Numerical investigation of the effect of vertical load on the lateral response of piles. Journal of Geotechnical and Geoenvironmental Engineering, 2007, 133: 512-521.

Kishore Y N, Rao S N, Mani J S. The behaviour of laterally loaded piles subjected to scour in marine environment. J. Civ. Eng., 2009, 13: 403-406.

Kühn M. Dynamics of offshore wind energy converters on monopile foundations-experience from the Lely offshore wind farm. OWEN Workshop "Structure and Foundations Design of Offshore Wind Turbines" March 1, 2000, Rutherford Appleton Lab.

Kuhn M. Offshore wind farms. In Wind Power Plants: Fundamentals, Design, Construction and Operation (Gasch R and Twele J (eds)). Springer, Heidelberg, Germany, 2002: 365-384.

Leblanc C, Houlsby G T, Byrne B W. Response of stiff piles in sand to long-term cyclic lateral loading. Géotech, 2010, 60: 79-90.

Li J L, Wang X F, Guo Y, et al. The loading behavior of innovative monopile foundations for offshore wind turbine based on centrifuge experiments. Renewable Energy, 2020, 152: 1109-1120.

Li J Z, Wang X L, Zhang H R. P-Y curve of weakening saturated clay under lateral cyclic load. China Offshore Platform, 2017, 32(3): 36-42. (In Chinese)

Li W, Igoe D, Gavin K. Evaluation of CPT-based p-y models for laterally loaded piles in siliceous sand. Geotechnique Letters, 2014, 4(2): 110-117.

Liang F Y, Qin C R, Chen S Q. Model test for dynamic p-y backbone curves of soil-pile interaction under cyclic lateral loading. China Harbour Engineering, 2017, 37(9): 21-26. (In Chinese)

Ling O, Wei X, Yue Q. Offshore wind zoning in China: Method and experience. Ocean and Coastal Management, 2018, 151: 99-108.

Liu H J, Zhang D D, Lv X H. A methodological study on the induction of triploidy oyster with different salinities. Periodical of Ocean University of China: Natural science edition, 2015, 1: 76-82. (In Chinese)

Liu J, Cui L, Zhu N, et al. Investigation of cyclic pile-sand interface weakening mechanism based on large-scale CNS cyclic direct shear tests. Ocean Engineering, 2019, 194: 106650.

Liu M M, Yang M, Wang H J. Bearing behavior of wide-shallow bucket foundation for offshore wind turbines in drained silty sand. Ocean Engineering, 2014, 82: 169-179.

Lombardi D, Bhattacharya S, Muir W D. Dynamic soil-structure interaction of monopile supported wind turbines in cohesive soil. Soil Dynamics and Earthquake Engineering, 2013, 49: 165-180.

Lombardi D, Bhattacharya S, Muir Wood D. Dynamic soil-structure interaction of monopile supported wind turbines in cohesive soil. Soil Dynamics and Earthquake Engineering, 2013, 49: 165-180.

Lowe J. Hornsea met mast-a demonstration of the "twisted jacket" design. Proceedings of the Global Offshore Wind Conference, ExCel London, London, UK, 2012.

Luengo J N, Vicente G B, Javier L G, et al. New detected uncertainties in the design of foundations for offshore Wind Turbines. Renewable Energy, 2018, 131: 667-677.

Ma H W, Yang J, Chen L Z. Numerical analysis of the long-term performance of offshore wind turbines supported by monopiles. Ocean Eng., 2017, 136: 94-105.

Martinez-Luengo M, Shafiee M, Kolios A. Data management for structural integrity assessment of offshore wind turbine support structures: data cleansing and missing data imputation. Ocean Engineering, 2019, 173: 867-883.

Matlock H. Correlations for design of laterally loaded piles in soft clay. Offshore technology conference, Houston, 1970, 1: 577-594.

Matlock H. Correlations for design of laterally loaded piles in soft clay. Proceedings of the 2nd Offshore Technology Conference, Houston, 1970: 22-24, 577-594.

Matsumoto T, Fukumura K, Kitiyodom P. Experimental and analytical study on behavior of model piled rafts in sand subjected to horizontal and moment load. International Journal of Physical Modeling in Geotechnical, 2004, 4: 1-19.

Mcaulty J F Thrust loading on piles. Journal of the Soil Mechanics and Foundations Division, 1956, 82: 1-25.

Meyerhof G G, Ghosh D P. Ultimate capacity of flexible piles under eccentric and inclined loads. Canadian Geotechnical Journal, 1989, 26: 34-42.

Meyerhof G G, Sastry V V R N. Bearing capacity of rigid piles under eccentric and inclined loads. Canadian Geotechnical Journal, 1984, 21: 267-276.

Mindlin R, Deresiewicz H. Elastic spheres in contact under varying oblique forces. ASME Journal of Applied Mechanics, 1953, 20: 327-344.

Murchinson J M, O'Neill M W. Evaluation of p-y relationships in cohesionless soils. Proceedings of Analysis and Design of Pile Foundations. San Francisco, 1984: 174-191.

O'Neill M W, Murchison J M. An evaluation of p-y relationships in sands. University of Houston, 1983.

O'Sullivan C, Cui L, O'Neil S. Discrete element analysis of the response of granular materials during cyclic loading. Soils and Foundations, 2008, 48(4): 511-530.

Perveen R, Kishor N, Mohanty S R. Offshore wind farm development: present status and challenges. Renew Sustain Energy Rev., 2014, 29: 78-92.

Poulos H G., Hull T S. The role of analytical geomechanics in foundation engineering. Foundation Engineering. ASCE, 1989: 1578-1606.

Reese L C, Cox W R, Koop F D. Analysis of laterally loaded piles in sand. Proceedings of the Offshore Technology Conference. Houston: University of Houston, 1974: 95-105.

Reese L C, Welch R C. Lateral loading of deep foundations in stiff clay. Journal of Geotechnical and Geoenvironmental Engineering, 1975, 101(7): 633-649.

Rosquoet F, Thorel L, Garnier J. Lateral cyclic loading of sand-installed piles. Soils and Foundations, 2007, 47(5): 821-832.

Sastry V V R N, Meyerhof G G. Behaviour of flexible piles under inclined loads. Canadian Geotechnical Journal, 1990, 27: 19-28.

Sastry V V R N, Meyerhof G G. Flexible piles in layered soil under eccentric and inclined loads.

Soils and Foundations, 1999, 39(1): 11-20.

Sawada K, Takemura J. Centrifuge model tests on piled raft foundation in sand subjected to lateral and moment loads. Soils and Foundations, 2014, 54: 126-140.

Sun Z Z, Zhang Y M, Yong Y, et al. Stability analysis of a fire-loaded shallow tunnel by means of a thermo-hydro-chemo-mechanical model and discontinuity layout optimization. International Journal for Numerical and Analytical Methods in Geomechanics, 2019: 1-14.

Vu A T, Matsumoto T. Experimental and numerical study on small-size piled raft foundation models subjected to cyclic horizontal loading. Geotechnics for Sustainable Infrastructure Development-Geotec Hanoi, 2016.

Wang J L. Numerical Study of Failure Mode of Pile Foundation for Offshore Wind Turbines. Tianjin University, Tianjin, China, 2012.

Wang X F, Zeng X W, Li J L. Vertical performance of suction bucket foundation for offshore wind turbines in sand. Ocean Engineering, 2019, 180: 40-48.

Wang X F, Zeng X W, Li X Y, et al. Liquefaction characteristics of offshore wind turbine with hybrid monopile foundation via centrifuge modelling. Renewable Energy, 2020, 145: 2358-2372.

Xiaoni W, Yu H, Ye L, et al. Foundations of offshore wind turbines: A review. Renewable and Sustainable Energy Reviews, 2019, 104: 379-393.

Xiong C W, Qi X, Gao A K, et al. The bypass transition mechanism of the Stokes boundary layer in the intermittently turbulent regime. Journal of Fluid Mechanics, 2020: 896.

Xu G M, Zhang W M. A study of size effect and boundary effect in centrifugal tests. Chin. J. Geotech. Eng., 1996, 18(3): 80-85. (In Chinese)

Yu L Q, Wang L Z, Guo Z, et al. Long-term dynamic behavior of monopile supported offshore wind turbines in sand. Theoretical and Applied Mechanics Letter, 2015, 5(2): 80-84.

Zhang C R, Yu J, Huang M S. P-Y curve analyses of rigid short piles subjected to lateral cyclic load in soft clay. Chin. J. Geotech. Eng., 2011, 33(S2): 78-82. (In Chinese)

Zhang L, Liu Y, Yan C. The model experimental study on characteristics of single pile hysteresis loops under horizontal cyclic loading. Shanxi Architecture, 2015, 41(11): 38-39. (In Chinese)

Zhang X L, Liu J X, Han Y, et al. A framework for evaluating the bearing capacity of offshore wind power foundation under complex loadings. Applied Ocean Research, 2018, 80: 66-78.

Zhang Y, Wang Z G, Zhao, S Z. Centrifugal tests of single pile's bearing capacity subjected to bidirectional cyclic lateral loading. Journal of Water Resources and Architectural Engineering, 2014, 12(4): 27-31. (In Chinese)

Zhu B, Xiong G, Liu J C. Centrifuge modelling of a large-diameter single pile under lateral loads in sand. Chin. J. Geotech. Eng., 2013, 35(10): 1807-1815. (In Chinese)

Zhu N, Cui L, Liu J W, et al. Discrete element simulation on the behavior of open-ended pile under cyclic lateral loading. Soil Dynamics and Earthquake Engineering, 2021, 144: 106646.

Zhu Z Y, Ren Q, Liu Y. Experimental reseach of stress distribution around single pile under long-time dynamic cyclic loading. Industrial Construction, 2017, 47(6): 102-107. (In Chinese)

Zhukov N V, Balov I L. Investigation of the effect of a vertical surcharge on horizontal displacement and resistance of pile columns to horizontal load. Soil Mechanics and Foundation Engineering, 1978, 15: 16-22.

Zormpa T E, Comodromos E M. Numerical evaluation of pile response under combined lateral and axial loading. Geotech. Geol. Eng, 2018, 36: 793-811.

Chapter 5 Investigation of cyclic pile-sand interface weakening mechanism

Under the complex wind and ocean wave-induced cyclic loads, the installed piles in pile-group foundations are prone to potential bearing capacity failures, which are usually initiated by pile-soil interface sliding due to interface material strength reduction (Randolph, 2003). However, this pile-soil interface weakening mechanism has not been thoroughly understood due to its dependency on the critical dynamic and geotechnical factors, such as the dynamic characteristics of cyclic loads, interface stress conditions, pile types, etc. Therefore, there is a need to investigate the cyclic pile-soil interface behavior in order to better understand the associated interface weakening mechanism and to account for the induced offshore pile foundation failures.

Most of the conventional pile-soil interface experiments apply the constant normal load (CNL) boundary conditions on soil specimens throughout shearing. This can simulate the external loading conditions of some engineering stability problems, such as anchored soil slopes and retaining walls under traffic loads. However, there are more practical problems, such as the pile foundations, where the normal load acting on the interface does not remain constant and therefore the CNL is not an appropriate representation of realistic boundary conditions. In this case, any volume change in the pile-soil interface zone is constrained by the soil beyond this zone. This can be simulated by applying a constant normal stiffness (CNS) boundary condition during the direct shear tests to represent the constraining effects from the surrounding soil. As a result, the normal stress on the interface may decrease or increase, depending on whether the soil in the interface zone contracts or dilates respectively. Therefore, the cyclic interface pile-soil behavior for pile foundation problems is usually investigated employing the CNS direct shear apparatus, such as in Johnston et al. (1987), Ooi and Carter (1987), Tabucanon et al. (1995), Porcina et al. (2003), Fakharian and Evgin (1997), Dejong et al. (2003, 2006), Jiang et al. (2004), Mortaraetal et al. (2007 and 2010), Peng et al. (2014) and Di Donna et al. (2016).

One of the limitations of the laboratory investigations for pile-soil interface behavior is the scaling effects. In other words, the effective shearing area of laboratory tests is considerably smaller than the prototype one. Therefore, in this chapter in order to reduce the influence the scaling effects and the boundary effects induced by shear box wall, a large-scale CNS cyclic direct shear apparatus is designed to investigate the pile-soil cyclic interface behavior. Investigations are concentrated in two common types of piles (concrete and steel) installed in the Fujian sand from South China Sea due to the significant development of offshore engineering in this area. The objectives of the work are to reveal the pile-sand interface cyclic weakening mechanism, in terms of the cyclic attenuation of shear/normal stress amplitude, strength reduction behavior and particle crushing of the interfacial material. Furthermore, the influences of initial confining pressure, cyclic shear deformation amplitude and pile surface roughness on pile-sand interface cyclic weakening mechanism are also investigated.

5.1 Large-scale CNS cyclic direct shear tests

5.1.1 Testing apparatus

As illustrated in Figure 5.1, the concerned domain for pile-soil interface investigations can be simplified into three zones, i.e. the shearing zone, the elastic zone and the undisturbed zone (White and Bolton, 2002; Liu et al., 2019). The cyclic shear deformation is mainly induced in the shearing zone with infinitesimal thickness compared to the pile dimension. Under cyclic loads, the soil particles in this zone can be re-arranged and crushed, resulting in plastic deformation and weakening effects on the interface soils. However, beyond the shearing zone, cyclic loads only induce recoverable elastic deformation in the soil adjacent to the interface. This zone is defined as the elastic zone and can be experimentally represented by the ideal constant normal stiffness boundary conditions (i.e. elastic springs) applied on the shear zone, as shown in Figure 5.1. Furthermore, in the far field, the induced deformation is insignificant, and this can be considered as the fixed end for the elastic springs, i.e. the undisturbed zone.

As shown in Figure 5.1, the spring stiffness k of elastic zone can be estimated by Equation $4G/D$ (Boulon and Foray, 1986), where D is the diameter of pile, and G is the operational shear modulus of the soil around the pile and may be expected to be about $0.4G_0$ (where G_0 is the undisturbed soil shear modulus) (Fahey et al., 2003).

As shown in Figure 5.2, the designed large-scale CNS direct shear apparatus consists of four main parts: the CNS loading system, the pile-soil interface direct shear system, the cyclic actuating system and the data acquisition system. The main components of the actuating and data acquisition systems are:

(1) The gear motor includes a frequency-variable motor and a speed reducer, with the following characteristics: power: 2kW; rotational speed: 1500r/min; frequency: 5-50Hz; speed reducer: 1/100; controlled shearing rate: 0.15-15mm/min; maximum shearing amplitude: 100mm.

(2) Fiber Bragg Grating (FBG) sensors are employed for measuring interface shear stresses, with the following characteristics: central wavelength: 1530nm; sensitivity: 3.7pm/με; range: ±2000με; resolution: 1με.

(3) A resistive earth pressure gauge is used for measuring interface normal stresses, with the following characteristics: diameter: 28mm; thickness: 10mm; range: 0-0.2MPa; sensitivity: 2.0; accuracy: $\leqslant 0.08\%$.

(4) An automatic data acquisition system.

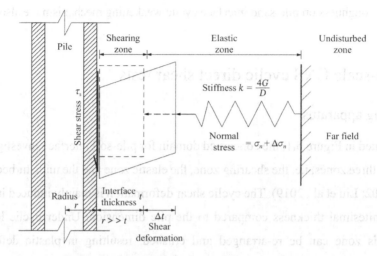

Figure 5.1 Schematic graph for CNS direct shear tests

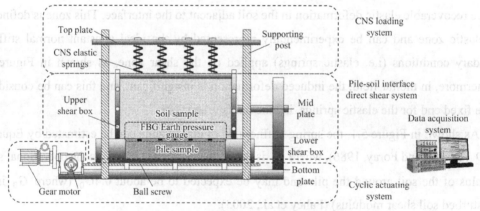

(a) Schematic diagram

(b) Physical apparatus

Figure 5.2 CNS direct shear apparatus

The layout of the FBG sensors and the earth pressure gauge, which are used to measure shear and normal stresses respectively, is shown in Figure 5.3. It is noted that the FBG sensors were sealed with epoxy resins, while the earth pressure gauge is in direct contact with the tested soil material. Furthermore, as shown in Figure 5.2 and Figure 5.3, the effective shearing area is 0.14m^2 and remains constant during shearing. This essentially mitigates the scaling effects involved in common laboratory interface direct shear tests and enhances the reliability of the experimental results in this study.

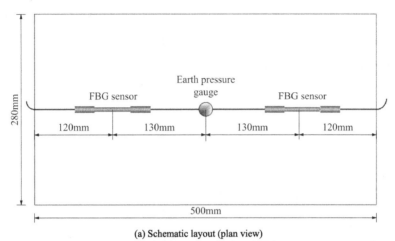

(a) Schematic layout (plan view)

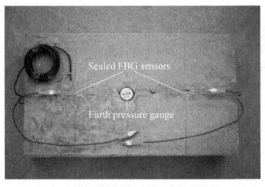

(b) Concrete pile (with sensors)

(c) Steel pile (with sensors)

Figure 5.3   The layout of FBG sensors and earth pressure gauge

5.1.2   Testing materials and testing scheme

The dry Fujian Sand was tested for the pile-soil interface investigations, as shown in Figure 5.4(a). The sand sample was taken from Xiamen in Fujian Province, southeast of China, where the offshore projects, such as wind turbines and oil and gas offshore platforms, are under significant development. The particle size distribution (PSD) and the material properties are shown in Figure 5.4(b) and Table 5.1 respectively.

For the tested two piles, the concrete pile was sampled using C40 cement with sand, gravel and water (the proportion is 2 : 4.1 : 7.8 : 1), following 28-day standard indoor cure, and the material of the steel pile was Q235 steel. It is noted that the surface roughness of the two piles were processed to fall in the range of 200-300μm and 40-50μm respectively, the same as the prototype values.

The laboratory tests were designed to investigate the effects of several critical factors on cyclic pile-sand interface behavior, i.e. the pile type (concrete and steel), shear deformation amplitude (5mm and 10mm) and initial confining pressure (90kPa and 110kPa). The adopted testing scenarios are listed in Table 5.2. It is noted that 20 cycles are relatively low when considering tidal loads. However, this research targeted its initial idea to firstly understand the interfacial cyclic weakening mechanism at the beginning of the large number cyclic loading. The pile-soil interfacial cyclic behavior under large cycle numbers will be considered in the future work.

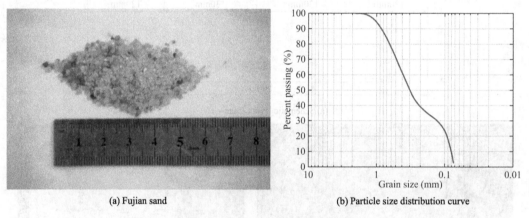

(a) Fujian sand (b) Particle size distribution curve

Figure 5.4   Testing soil material

Material properties   Table 5.1

Specific weight of Soil particle G_s	Maximum void ratio e_{min}	Minimum void ratio e_{max}	Average particle size D_{50} (mm)	Relative density D_r (tested sample)	Internal friction angle φ (°)
2.71	0.40	0.86	0.34	90%	40.4

Testing scheme   Table 5.2

Test number	Initial confining pressure (kPa)	Shear deformation amplitude (mm)	Pile type	Shearing rate (mm/min)	Cycles
CS-1	90	10	Concrete-sand (CS)	5	20
CS-2	110	10	Concrete-sand (CS)	5	20
CS-3	90	5	Concrete-sand (CS)	5	20
SS-1	90	10	Steel-sand (SS)	5	20
SS-2	110	10	Steel-sand (SS)	5	20
SS-3	90	5	Steel-sand (SS)	5	20

5.2 Monotonic pile-sand interface behavior

Monotonic tests were first carried out to investigate the pile-sand interface behavior, which can also be considered as the validation for the designed CNS direct shear apparatus. Figure 5.5 shows the shear stress-displacement behavior at the pile-sand interface of the two tested piles under two initial confining pressures. Overall, the tested sand shows a softening behavior after reaching the peak strength, which then gradually approaches the residual strength as the shear deformation increases. In particular, the concrete-sand reaches the peak strength at the shear displacements of 3.3mm and 3.1mm for the confining pressures of 90kPa and 110kPa respectively. The steel-sand shows a faster trend reaching the peak strength, at the displacements of 1.7mm and 1.6mm for the two confining pressures. It can be seen that the strength of the concrete-sand interface is larger than that of the steel-sand interface, showing the positive effects of the higher surface roughness of the concrete pile on interface shear resistance. Furthermore, the higher confining pressure induces larger interface strength, and this influence is more significant for rougher piles. These observations are in agreement with the monotonic results from Fakharian and Evgin (1997), showing the reliability of the designed CNS direct shear apparatus for the subsequent cyclic investigations. It is noted that the experiments were conducted under two confining pressure conditions: 90kPa and 110kPa. Figure 5.5 presents both results to show the effects of confining pressure on the cyclic interfacial behavior. It is noted that the normal stress we used in each test is constant.

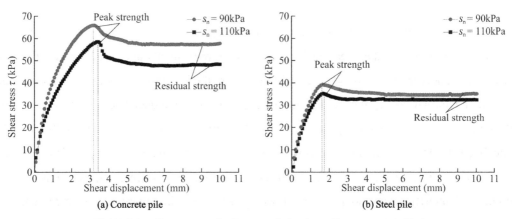

Figure 5.5   Shear stress-displacement behavior under monotonic load

5.3 Cyclic pile-sand interface weakening mechanism

In this section, the cyclic pile-sand interface behaviors of the concrete and steel piles are investigated under different confining pressures and cyclic deformation amplitudes. The experimentally obtained results, including the induced shear/normal stress-displacement behavior, stress path behavior and particle size distribution at the pile-sand interface, are analyzed. This further leads to a detailed study for the associated pile-sand interface cyclic weakening mechanism, through the investigations of cyclic attenuation of shear/normal stress amplitude, strength reduction and particle crushing, respectively.

5.3.1 Cyclic attenuation of interface shear and normal stress amplitudes

The experimentally obtained interface stress-displacement behavior for the concrete pile is shown in Figure 5.6, under two confining pressures and subjected to two cyclic deformation amplitudes.

Taking the results from the CS-1 test for example (Figures 5.6(a) and (b)), overall, both the shear and normal stress amplitudes tend to attenuate as cycle number increases. In particular, for the shear stress-displacement behavior (Figure 5.6(a)), under the initial loading of the first cycle, the interface sand reaches the peak shear stress of 59kPa at the displacement level of 3.3mm, and gradually approaches the residual strength of 48kPa, which is the same as the monotonic behavior discussed in the previous section. However, compared to the residual strength at the 1st cycle, the residual strength reduces as the cycle number increases, with the reduction ratio of 34%, 42%, 48% and 52% at the 5th, 10th, 15th and 20th cycle respectively. It can be seen that the attenuation is the most significant in the first 5 cycles and the shear stress amplitude gradually stabilizes after 15 cycles.

As seen from Figure 5.6(b), the normal stress amplitude also attenuates as cycle number increases. This is due to the shearing induced particle movements and crushing and thus soil densification under the CNS boundary conditions, where the interface zone thickness reduces and therefore the normal load applied on the interface is relaxed. The induced normal stress amplitude reduction ratios at the maximum positive shear deformation levels are 30%, 40%, 48% and 52% at the 5th, 10th, 15th and 20th cycle respectively. Again, the normal stress amplitude attenuation is the most significant in the first 5 cycles. It is noted that the normal stress-displacement behavior presents a "butterfly-shape" loop. In other words, as shown in Figure 5.6(b), under both the positive-directional loading (④①) and negative-directional loading (②③), the normal stress decreases first and then increases. This reflects that the sand in the interface zone contracts first and then dilates. Similar results were also observed by Cui and Bhattacharya (2016) in the numerical simulations of cyclic soil-pile interaction using Discrete Element Modeling. Consequently, for the conducted tests under the CNS boundary conditions, the induced normal load decreases and increases. However, the observed alternating particle contraction and dilation in this work can only reflect the cyclic interfacial behavior specifically for the investigated sand. Some investigations showed particle contraction and stress relaxation behavior only during cyclic interface direct shear tests (Silver, 1971).

The stress-displacement behaviors subjected to other confining pressure and shear deformation amplitude are shown in Figures 5.6(c)-(f), for the CS-2 and CS-3 tests. Similar patterns for the stress-displacement behavior are observed from these two tests, compared to those for the CS-1 test. In particular, hysteretic and butterfly loops can be found for the shear stress and normal stress behavior respectively, with amplitudes cyclically attenuating as cycle number increases.

In order to investigate the influence of confining pressure and shear deformation amplitude on the observed attenuation behavior, the obtained shear and normal stress amplitudes are plotted against cycle numbers in Figure 5.7. Based on Liu et al. (2012), the attenuation trends of the shear and normal stress amplitudes can be best fitted with the logarithmic models expressed by Equations (5.1) and (5.2) respectively.

$$\tau = m - n\ln(N - l) \tag{5.1}$$

$$\sigma = a - b\ln(N - c) \tag{5.2}$$

where parameters m and a represent the initial shear and normal stress; parameters n and b are the factors which quantify the stress amplitude attenuation rate; N is the cycle number; l and c are considered as the factors which influence the locations of curves. All the model parameters which can be obtained by curve fitting of the experimental results.

Based on Figure 5.7, overall, the attenuations of the shear and normal stress amplitudes are most significant in the first 5 cycles, and the degradation rate gradually decreases as cycle number increases and is negligible after 15 cycles. The comparison of the results between the CS-2 and CS-1 tests show that a higher confining pressure would contribute to a higher shear strength on the pile-sand interface. Furthermore, by comparing the values of the parameter b in the best-fit attenuation curves for the CS-2 and CS-1 tests, the higher confining pressure induces a more significant attenuation behavior for the shear stress amplitude. In particular, under higher normal loads, soil particles at the pile-sand interface can be crushed more significantly, which therefore induces a more profound attenuation of the shear stress amplitude. This is also in agreement with the comparison of the normal stress attenuation between the CS-2 and CS-1 results. When comparing the results between the CS-3 and CS-1 tests, it can be seen that larger shear deformation amplitude leads to a lower interface shear stress and slightly more significant attenuation behavior.

Pile foundation for offshore wind power: macro and micro mechanism of load bearing performance

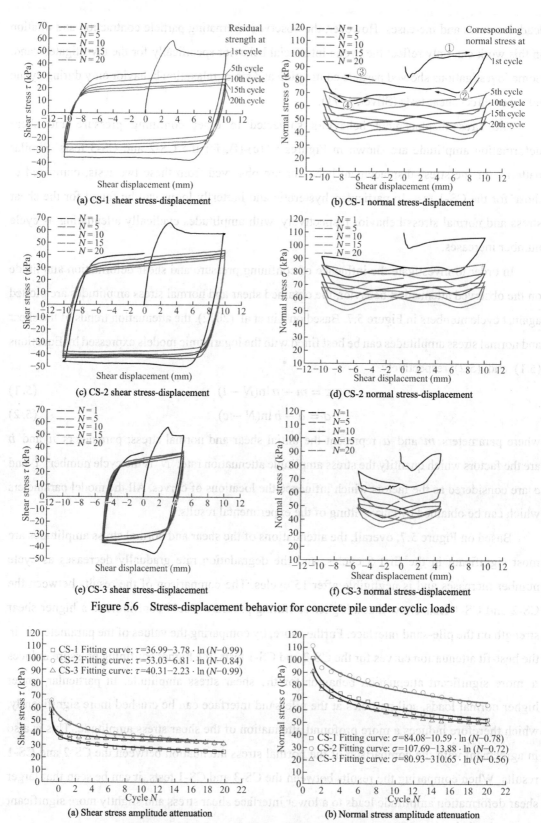

Figure 5.6 Stress-displacement behavior for concrete pile under cyclic loads

(a) Shear stress amplitude attenuation

(b) Normal stress amplitude attenuation

Figure 5.7 Stress amplitude attenuation of the concrete pile

The interface stress-displacement behaviors for the steel pile under two confining pressures and subjected to two cyclic deformation amplitudes are shown in Figure 5.8. The overall shear and normal stress-displacement behaviors are similar to those of the concrete pile. In particular, hysteretic and butterfly loops can be found for the shear stress and normal stress behavior respectively, with amplitudes cyclically attenuating as cycle number increases. However, for the normal stress behavior, the alternating contractive and dilative behaviors are not as significant as those at the concrete-sand interface. This is mainly due to the higher roughness of concrete surface, which causes more significant particle movements, thus more profound dilation or contraction. Furthermore, the overall shear strength amplitude (i.e. the height of the hysteretic loop) is smaller than that of the concrete pile. This is also reflected by comparing the shear stress amplitude attenuation curves in Figure 5.9 with those in Figure 5.7. This is mainly induced by the lower surface roughness at the steel-sand interface compared to the concrete pile (approximately 40-50μm and 200-300μm respectively), resulting in lower friction angle and therefore fewer particle rearrangements and less energy dissipation. Furthermore, it is found that the attenuation rate of the shear stress amplitude (quantified by the parameter n) is larger under higher confining pressure and larger shear deformation amplitude, which is in agreement with the observations from the concrete-sand tests.

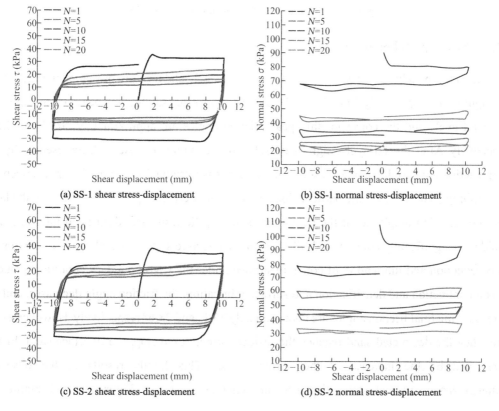

(a) SS-1 shear stress-displacement

(b) SS-1 normal stress-displacement

(c) SS-2 shear stress-displacement

(d) SS-2 normal stress-displacement

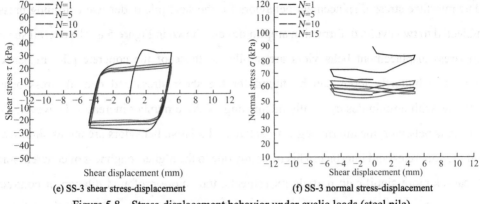

(e) SS-3 shear stress-displacement (f) SS-3 normal stress-displacement

Figure 5.8 Stress-displacement behavior under cyclic loads (steel pile)

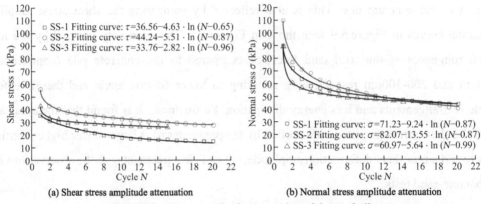

(a) Shear stress amplitude attenuation (b) Normal stress amplitude attenuation

Figure 5.9 Stress amplitude attenuation of the steel pile

5.3.2 Stress path behavior

The stress path behavior of the concrete-sand and steel-sand tests are presented in Figure 5.10 and Figure 5.11 respectively. For CS-1 and CS-2 tests, the tested sands reach peak strength envelop at the 1st cycle and the corresponding peak friction angles are approximately 30.9° and 32.6° respectively. After the 1st cycle, significant relaxation behavior is observed, represented by the sharp friction angle reduction. This softening behavior is attributed to the particle rearrangement and crushing (especially for the particles locked in rough interface asperities) induced by the large shear deformation amplitude at the beginning of the cyclic tests. The destructed interface sand would then significantly contract and fill gaps on the concrete surface, which leads to the normal stress reduction and lubrication of concrete surface under the CNS boundary conditions. Due to the soil contraction and densification and surface lubrication, the friction angle decreases, and the corresponding final friction angles are approximately 26.1° and 28.6° respectively. However, every time when the destructed sand reaches the residual strength envelop, the interface sand dilates, represented by the increase of the normal stress. This finally results in the observed "butterfly-shape" normal stress-displacement behavior shown in Figure 5.6 and Figure 5.8.

Increase in the shear deformation amplitude produced larger loops. However, for the CS-3 test, the difference between the peak and final friction angles is insignificant, due to the small shear deformation amplitude applied in the CS-3 test.

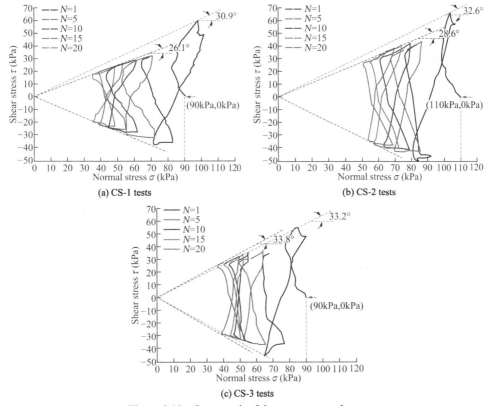

Figure 5.10 Stress path of the concrete-sand tests

Regarding the stress path behavior obtained from the steel-sand tests, a slight hardening behavior is evident in Figure 5.11, with a higher friction angle being mobilized at lower normal stresses. In particular, the friction angles for the three steel-sand tests, are 23.3°, 22.4° and 23.1° at the first cycle, while reduce to the residual strength of 30.9°, 29.8° and 29.5° respectively. This phenomenon was also observed by Dejong et al. (2006) in soil-structure interface direct shear tests. This increase in friction angle (i.e. hardening behavior) can be attributed to the reduction of normal stress and increase of density (Boulon and Foray, 1986). Another difference observed for the steel interface, compared to the concrete interface, is that no loops are observed for the first cycle. However, with the increase of relative density during the following cycles, loops occur. Similar patterns were observed by Mortara et al. (2007).

The experimentally obtained peak interface friction angles of all the tests are listed in Table 5.3. The ratio between the peak interface friction angles and natural sand internal friction angles are in the range of 0.73-0.81. This ratio is found to be smaller for the steel pile, which results in the lower interface shear strength, in agreement with the observations in the previous section. It is noted that when carrying out the safety design and analysis for offshore pile foundation, the interfacial friction angle is usually assumed to be constant, as friction angle is determined by the pile material, soil material and soil relative density. However, this assumption only represents the initial condition, as interface soil is subject to densification under cyclic shearing. This can be justified by the results shown in Figure 5.10 and Figure 5.11 and Table 5.3.

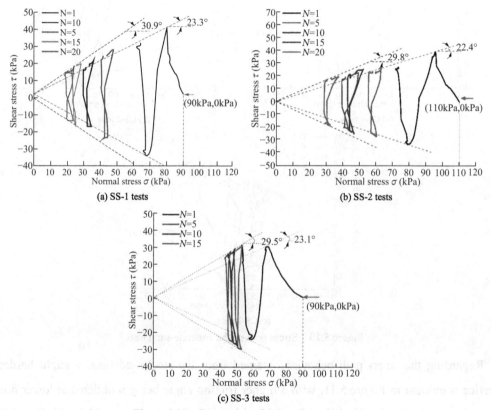

Figure 5.11 Stress path of the steel-sand tests

Interface friction angles and angles of shearing resistance   Table 5.3

Tests	CS-1	CS-2	CS-3	SS-1	SS-2	SS-3
Peak interface friction angle $\varphi_{interface}$ (°)	30.9	32.6	33.2	30.9	29.8	29.5
$\varphi_{interface}/\varphi$	0.76	0.81	0.74	0.76	0.74	0.73

5.3.3 Interface particle crushing

In order to more specifically investigate the pile-sand interface weakening mechanism from micro perspective, the particle size distribution of the sands at different distances from the interface (i.e. 0-1cm, 1-2cm, 2-3cm, 3-4cm and 4-5cm from the interface) after the cyclic tests were experimentally obtained and are shown in Figure 5.12 and Figure 5.13 for the concrete-sand and steel-sand tests respectively. Furthermore, based on the particle size distribution curves, the d_{50} at the different locations was obtained and plotted against the distance from the interface in Figure 5.14 for all the tests. Overall, it is observed that the particle crushing is only significant within 1cm from the interface and is negligible beyond this range. The cyclically induced particle crushing is found to be more profound under higher confining pressure (by comparing the CS-1/SS-1 and CS-2/SS-2 results) or subjected to larger shear deformation amplitude (by comparing the CS-1/SS-1 and CS-3/SS-3 results). This induces more significant stress amplitude attenuation under these two conditions as observed in Figure 5.7. Furthermore, it shows that the particle crushing is more significant at the concrete-sand interface, due to the higher surface roughness of the concrete pile.

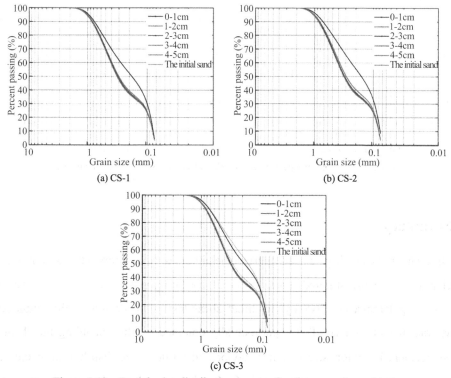

Figure 5.12 Particle size distribution curves after the concrete-sand tests

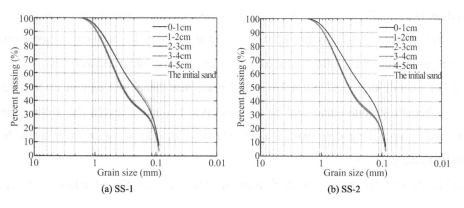

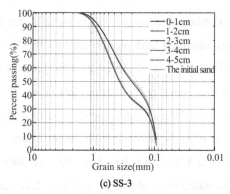
(c) SS-3

Figure 5.13 Particle size distribution curves after the steel-sand tests

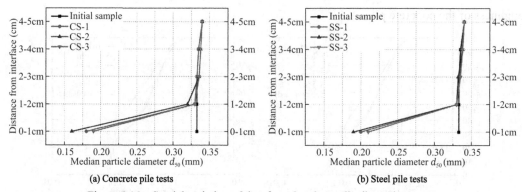

(a) Concrete pile tests   (b) Steel pile tests

Figure 5.14 Spatial variation of the d_{50} after the cyclic direct shear tests

5.4 Summary

In this chapter, the cyclic pile-sand interface behaviors of two common types of piles (concrete and steel piles) were investigated under different confining pressures and cyclic deformation amplitudes using large-scale CNS cyclic direct shear tests. The design of the employed CNS apparatus was first introduced. The experimentally obtained results, including the shear/normal stress-displacement behavior, stress path behavior and particle size distribution at the pile-sand interface, were subsequently analyzed. This further led to a detailed study for the associated pile-sand interface cyclic weakening mechanism through the investigations of cyclic attenuation of shear/normal stress, strength reduction and micro-scale particle crushing, respectively. Key Findings regarding the cyclic pile-sand interface behavior are summarized as below:

(1) The overall shear and normal stress-displacement behaviors exhibit as hysteretic and butterfly (alternating contractive and dilative behavior) loops respectively. Higher confining pressure and smaller shear deformation amplitude contribute to a higher shear strength on the pile-sand interface.

(2) The shear and normal stress amplitudes attenuate significantly in the first 5 cycles, and the

degradation rate gradually becomes smaller and is negligible after 15 cycles. The reduction rate is more profound under higher initial confining pressure and larger shear deformation amplitude.

(3) Stress path results showed that under larger shear deformation amplitude the sand at concrete-pile interface tends to reach the peak strength at the beginning of the cyclic loading and then gradually reduces to the residual strength, showing a softening behavior. This is due to the significant particle crushing at the 1st cycle induced by the large shear deformation. However, for the steel-sand tests, a slight hardening behavior is evident, with a higher friction angle being mobilized at lower normal stresses.

(4) PSD investigations showed that, cyclically induced particle crushing is only significant within 1cm from the interface and is negligible beyond this range. Particle crushing was found to be more profound under higher confining pressure and subjected to larger shear deformation amplitude. Furthermore, it showed that the particle crushing is more significant at the concrete-sand interface compared to its counterpart, due to the higher surface roughness of the concrete pile.

It is noted that only dry sands were considered for this work. However, this manuscript is the first stage of our long-term research project. A clear understanding about the pile-sand interface cyclic behavior in the situation we presented in this manuscript is necessary before we further consider the offshore environment.

References

Altuhafi F N, Jardine R J, Georgiannou V N, et al. Effects of particle breakage and stress reversal on the behaviour of sand around displacement piles. Geotechnique, 2018, 68(6): 546-555.

Boulon M, Foray P. Physical and numerical simulation of lateral shaft friction along offshore piles in sand. Third International Conference on Numerical Method in Offshore Piling, Editions Technip, France, 1986: 127-147.

Buckley R, Jardine R J, Kontoe S, et al. Ageing and cyclic behaviour of axially loaded piles driven in chalk. Géotechnique, 2018, 68(2): 146-161.

Chow F C. Investigations into Displacement Pile Behaviour for Offshore Foundations. Ph.D. Thesis. Imperial College, University of London, 1997.

Cui L, Bhattacharya S. Soil-monopile interactions for offshore wind turbines. Engineering and Computational Mechanics, 2016, 169(4): 171-182.

Dejong J T, Randolph M F, White D J. Interface load transfer degradation during cyclic loading: a microscale investigation. Soils Found, 2003, 43(4), 91-94.

Dejong J T, White D J, Randolph M F. Microscale observation and modelling of soil-structure interface behaviour using PIV. Soils Found, 2006, 46(1): 15-28.

Di Donna A, Ferrari A, Laloui L. Experimental investigations of the soil-concrete interface: physical mechanisms, cyclic mobilization, and behaviour at different temperatures. Can Geotech J, 2016, 53(4): 659-672.

Fahey M, Lehane B M, Stewart D. Soil stiffness for shallow foundation design in the Perth CBD. Australian Geomechanics Journal, 2003, 38(3): 61-90.

Fakharian K, Evgin E. Cyclic simple-shear behavior of sand-steel interfaces under constant normal stiffness condition. J Geotech Geoenviron, 1997, 123(12): 1096-1105.

Huang M S, Liu Y. Axial capacity degradation of single piles in soft clay under cyclic loading. Soils Found, 2015, 55(2): 315-328.

Jardine R J, Standing J R. Field axial cyclic loading experiments on piles driven in sand. Soils Found, 2012, 52(4): 723-736.

Jiang Y, Xiao J, Tanabashi Y, et al. Development of an automated servo-controlled direct shear apparatus applying a constant normal stiffness condition. Int J Rock Mech Min, 2004, 41(2): 275-286.

Johnston I W, Lam, T S K, Williams A F. Constant normal stiffness direct shear testing for socketed pile design in weak rock. Géotechnique, 1987, 37(1): 83-89.

Li Z, Bolton M D, Stuart, K H. Cyclic axial behaviour of piles and pile groups in sand. Can Geotech J, 2012, 49(9): 1074-1087.

Liu J W, Duan N, Cui L, et al. DEM investigation of installation responses of jacked open-ended piles. Acta Geotechnica, 2019. DOI: 10.1007/s11440-019-00817-7.

Liu J W, Zhang Z Z, Yu F. Case history of installing instrumented jacked open-ended piles. J Geotech Geoenviron, 2012, 12(7): 810-820.

Lombardi D, Bhattacharya S, Wood D M. Dynamic soil-structure interaction of monopile supported wind turbines in cohesive soil. Soil Dyn Earthq Eng, 2013, 49(18): 165-180.

Mortara G, Ferrara D, Fotia G. Simple model for the cyclic behaviour of smooth sand-steel interfaces. J Geotech Geoenviron, 2010, 136(7): 1004-1009.

Mortara G, Mangiola A, Ghionna V N. Cyclic shear stress degradation and post-cyclic behaviour from sand-steel interface direct shear tests. Can Geotech J, 2007, 44(7): 739-752.

Ooi L H, Carter J P. A constant normal stiffness direct shear device for static and cyclic loading. Geotech Test J, 1987, 10(1): 3-12.

Peng S Y, Ng C W W, Zheng G. The dilatant behaviour of sand-pile interface subjected to loading and stress relief. Acta Geotech, 2014, 9(3): 425-437.

Porcino D, Fioravante V, Ghionna V N, et al. Interface behavior of sands from constant normal stiffness direct shear tests. Geotech Test J, 2003, 26(3): 1-13.

Randolph M F. Science and empiricism in pile foundation design. Geotechnique, 2003, 53(10): 847-875.

Rimoy S P. Ageing and axial cyclic loading studies of displacement piles in sands. Ph.D. Thesis. Imperial College, University of London, 2013.

Silver M L. Volume change in sands during cyclic loading. Proceedings of ASCE Journal of the Soil Mechanics & Foundation Division, 1971: 1171-1182.

Standing J R, Jardine R J, Rimoy S P. Displacement response to axial cycling of piles driven in sand. ICE Geotechnical Engineering, 2013, 166(2): 131-146.

Tabucanon J T, Airey D W, Poulos H G. Pile skin friction in sands from constant normal stiffness tests. Geotech Test J, 1995, 18(3): 350-364.

Tsuhaa C H C, Forayb P Y, Jardine, R J, et al. Behaviour of displacement piles in sand under cyclic axial loading. Soils Found, 2012, 52(3): 393-410.

White D J, Bolton M D. Displacement and strain paths during plane strain model pile installation in sand. Géotechnique, 2002, 54(6): 375-398.

Zhang B J, Mei C, Huang B, et al. Model tests on bearing capacity and accumulated settlement of a single pile in simulated soft rock under axial cyclic loading. Geomechanics and Engineering, 2017, 12(4): 611-624.

Zhou W J, Wang L Z, Guo Z, et al. A novel t-z model to predict the pile responses under axial cyclic loadings. Computers and Geotechnics, 2019, 112: 120-134.

Chapter 6 Bearing characteristics of wind turbine jacket group pile foundation in sandy soil

6.1 Model tests of jacking installation and lateral loading performance of jacket foundation in sand

Jacket supported OWTs are relatively new structures with few field installation observations and service records, while they are bound to operate for about 30 years. Moreover, the first-order natural frequency of an OWT is generally close to the forcing frequency applied by environmental loads. Since the OWTs are dynamically sensitive structures, the small change of their natural frequency can easily lead to resonance failure. During their whole service period, jacket supported OWT structures are subjected to approximately 10^7-10^8 cycles of loadings. It has been known that load transfer mechanism and foundation stiffness (natural frequency) will be changed under long-term cyclic loadings. Therefore, reliable installation data of the OWT foundations and accurate prediction of its long-term dynamic performance are very essential.

There have been few studies so far on the installation characteristics of jacket foundation, and most of them focused on the static/dynamic loading tests of pile foundation after installation. Yan et al. (2020) conducted large-scale laboratory model tests to explore jacking installations and lateral loading of a new skirted foundation in sand. Kou et al. (2019) presented model tests of steel caisson during and after installation with different penetration velocities in medium dense sand. It was found that the jacking force for caisson was significantly affected by the penetration velocity. Amaral et al. (2020) investigated the characterization of impact pile driving signals during installation of offshore wind turbine foundations. Heins et al. (2018) analyzed the effect of installation method on static and dynamic load test response for piles in sand. Lian et al. (2014) and Chen et al. (2016) conducted a series of laboratory tests to explore the interactions between the caisson and silty sand in both jacking and suction installation processes. Lee et al. (2015) investigated the vertical, horizontal and cone tip resistances of bucket foundations embedded in sand with different installation methods by conducting model tests. Koteras and Ibsen (2019) presented medium-scale tests to compare jacking and suction installations. Zhu et al. (2020) conducted model tests to study the penetration and tip resistance and earth pressure of bucket foundations embedded in sand with the jacking installation method. Zhu et al. (2011) carried out a series of large-scale model tests on suction installation and lateral loading of caisson foundations in saturated silt.

For foundations supporting OWTs, the vertical load imposed by self-weight is much lower than the vertical capacity. However, the lateral loads, including horizontal load and moment caused by winds, waves and currents, are more prone to lead to capacity failure or over-limit deformation of OWTs. Compared with the few studies on the jacket foundation, there has been abundant research on the lateral performance of monopile and suction caisson foundations in sand, with well-scaled model tests at 1g or in the centrifuge and numerical modeling (Liu et al., 2020; Abadie et al., 2019; Liu et al., 2019; Ma et al., 2019; Wang et al., 2018; Zhu et al., 2018; Duan et al., 2017; Wang et al., 2017; Depina et al., 2015; Choo et al., 2014; Klinkvort and Hededal, 2013; Zhu et al., 2013; Leblanc et al., 2010; Achmus et al., 2009). Additionally, a lot of investigations have been done to explicit dynamic performance of OWTs such as natural frequency and system damping. He and Zhu (2019) conducted 1g model tests in dry sand to investigate the frequency responses of the monopile under different loading conditions. Futai et al. (2018) presented a series of centrifuge tests for the first time directly to determine the natural frequency of the pile-soil system. Bisoi and Haldar (2018) investigated the long-term performance of monopile in clay using a series of scaled model tests. Hanssen et al. (2016) conducted a 1∶20 model scale test of a monopile foundation in dry sand to analyze the system eigenfrequency and damping. Carswell et al. (2015) used a linear elastic two-dimensional finite element model to appraise the significance of foundation damping on monopile-supported OWTs subjected to extreme storm loading. Guo et al. (2015) conducted a series of well-scaled model tests to investigate the long-term performance of the OWT founded on a monopile in dense sand. Lombardi et al. (2013) presented a series of scaled model tests in kaolin clay and the changes in natural frequency and damping of the model were monitored. In the recent the focus of studies about jacket foundations supporting OWTs has been mainly concentrated on its bearing characteristics. Zhu et al. (2019, 2018) presented a series of centrifuge tests and numerical studies to investigate lateral loading behavior of tetrapod piled jacket foundations in sand and clay. Kong et al. (2019) conducted centrifuge model tests on the cyclic lateral behavior of a tetrapod piled jacket foundation in medium dense sand.

Different from the previous studies, the installation characteristics and lateral loading performance of jacket foundations supporting OWTs are presented and discussed in this chapter. The jacking installation, static and cyclic lateral loading tests of the OWT model were carried out by using the large-scale indoor model test system. Moreover, the influence of load frequency and cyclic load ratio on the dynamic response of jacket foundation is analyzed. It is indicated that loading conditions have a significant effect on the dynamic behavior of the jacket supported OWT, which provides a new insight into the long-term performance of the OWT prototype. The results from this study will contribute to the guidance and prediction for the installation characteristics and long-term dynamic performance of jacket supported OWT structures.

6.1.1 Model testing

6.1.1.1 Equipment and materials

Model tests for the jacket supported OWT were carried out by using the large-scale indoor model test system. The test system was composed of five parts: model box, pile jacking system, horizontal static loading system, horizontal cyclic loading system and data acquisition system. The boundary effect can be eliminated by the model box with the dimensions of the 3m × 3m × 2m (length × width × height). A hydraulic jack was fixed on the top steel beam to provide reaction force used in jacking installation tests. Fixed pulley, elastic rope and mass block were used to make up a novel horizontal static loading system. An innovative horizontal cyclic loading device was composed of a plate, two gears, masses and a DC motor.

The experiments employed natural marine sand collected from coastal areas of Qingdao, China. The basic physical parameters of the sand are listed in Table 6.1. Furthermore, the particle grading curve is shown in Figure 6.1. According to the drainage condition of saturated sand around pile during the service period of OWT, the naturally air-dried sand was used for model tests (Zhang et al., 2019).

Basic physical parameters of Qingdao Sea sand   Table 6.1

Physical parameters	Value
Relative density G_s	2.65
Maximum void ratio e_{max}	0.52
Minimum void ratio e_{min}	0.30
Relative compaction D_r	0.73
Average particle size d_{50} (mm)	0.72
Particle size range (mm)	0-15
Internal friction angle (°)	42.8
Dry density (kg/m³)	1950
Shear modulus (MPa)	12.45

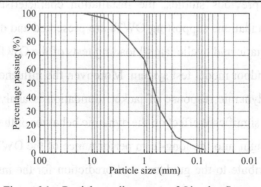

Figure 6.1   Particle grading curve of Qingdao Sea sand

The detailed physical parameters of jacket supported OWT model, which were determined for considering a scaling ratio of 1∶50, are listed in Table 6.2. The size and mass of each part were ascertained according to dimensionless similarity theory (Bhattacharya et al., 2011). The experimental setup and sensor placement are shown in Figure 6.2. The static loading height was 1.83m, and the cyclic loading height was 1.4m (Huang, 2017). The installation position and loading height of model are displayed in Figure 6.3.

Boundary effect can be negligible by controlling the distance between pile and the inner wall of model box to be greater than $10D$ (D is the diameter of pile), as shown in Figure 6.3 (Zhang et al., 2019). The jacket pile foundation was composed of four identical closed pipe piles. The pile diameter and length were 50mm and 1000mm, respectively. Furthermore, the wall thickness of pile was 1.5mm. Each pile was symmetrically arranged with 9 pairs of strain gauges on the outer wall. The specific distribution of strain gauges is shown in Figure 6.4.

The size and mass of each part of the model　　Table 6.2

Part	Length (mm)	Diameter (mm)	Wall thickness (mm)	Mass (kg)
Top mass block	/	/	/	2.4
Tower	1300	80	4	4.7
Jacket platform	530 (height)	/	/	3.32
Model pile	1050	50	1.5	3.32

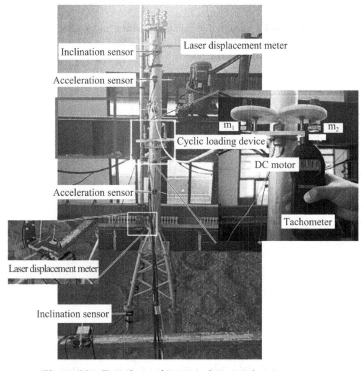

Figure 6.2　Experimental setup and sensor placement

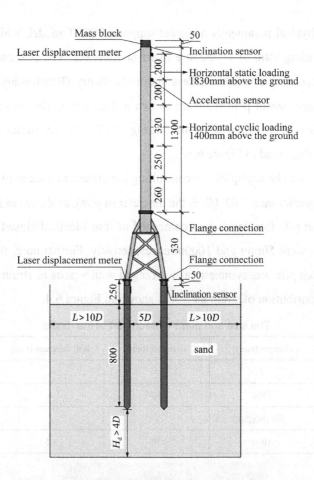

Figure 6.3 Sketch of installation position and loading height

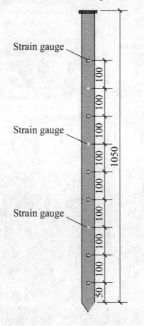

Figure 6.4 Distribution diagram of strain gauge (Unit: mm)

6.1.1.2 Testing procedure

Five tests were carried out in this study (T1 to T5), which were classified into three basic parts. The first part was jacking installation of pile foundation. The second part was static loading on tower, and the third one was lateral cyclic loading. Efforts were made to keep each testing condition unchanged.

Installation

The sand was poured into the model box in layers to control its relative density at about 73%. In order to control the verticality of pile, leveling staff was used to observe the inclination of installation. Four piles were first sunk into the sand under their self-weights. Then, Pile 1 to Pile 4 were jacked more deeply in sequence using hydraulic cylinder. During jacking process, strains of piles were obtained by data acquisition system. Finally, the penetration depth of pile was 800mm and the pile spacing was 300mm. The layout of pile jacking is shown in Figure 6.5.

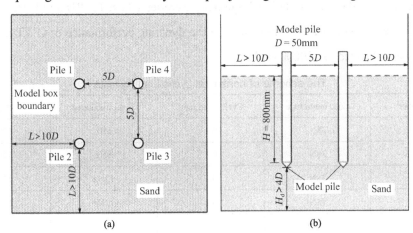

Figure 6.5 Layout of pile jacking
(a) Elevation view; (b) Side view

Lateral static loading

It is necessary to determine the lateral bearing capacity of jacket foundation before lateral cyclic loading. Monotonic lateral loads were applied through stepwise loading until the foundation failed. The time interval between each load level was ten minutes. Laser displacement meter and inclination sensor were employed to capture the displacement at the tower top and the tilt angle at the ground, respectively. According to the displacement development of tower top, the ultimate horizontal capacity of OWT model was determined to be 380N. The calculation method of ultimate capacity is illustrated in Figure 6.6.

Lateral cyclic loading

In order to investigate the effect of load conditions including load magnitude and load

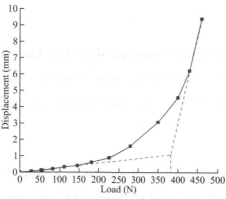

Figure 6.6 Load-displacement curve of static loading

frequency on the long-term behaviors of jacket supported OWT model, five experiments were conducted with varying factors as listed in Table 6.3. The two-way horizontal cyclic loading was provided by the innovative gear device. The form of cyclic load was a sine wave with the load peaks respectively chosen as 38N, 76N and 114N. Three load amplitudes corresponded to 10%, 20% and 30% of the ultimate horizontal capacity. In other words, the cyclic load ratios were 0.1, 0.2 and 0.3 respectively. It was determined that the first mode natural frequency of OWT model was 10Hz in free vibration tests. Due to the fact that the ratio of excitation frequency to natural frequency of OWT was close to one in the field (Bhattacharya et al., 2011), the excitation frequencies were set to be 6Hz, 8Hz and 12Hz respectively to investigate the influence of load frequency on the dynamic performance of OWT model. Each experiment was subjected to 120000 cycles.

The scheme of horizontal cyclic loading tests   Table 6.3

Test number	Load magnitude	Cyclic load ratio	Load frequency	Cycle number
T1	38N	0.1	6Hz	120000
T2	38N	0.1	8Hz	120000
T3	38N	0.1	12Hz	120000
T4	76N	0.2	12Hz	120000
T5	114N	0.3	12Hz	120000

6.1.2 Test results and discussions

6.1.2.1 Jacking installation

Installation resistance

Due to the limitation of sensor arrangement, the axial force calculated from the strain gauge data at the bottom of pile is regarded as the pile tip resistance. The variation curves of pile tip resistance, frictional resistance and total pile jacking resistance are shown in Figure 6.7. The curves of pile tip resistance are generally in accordance with each other. It is indicated that the linear growth of pile tip resistance can be divided into two stages: rapid growth stage and slow growth stage. The growth rate is large in the early stage of pile penetration and diminishes in the later stage, demonstrating that there is a critical depth of pile tip resistance. Corresponding to penetration depth 550mm, the pile tip resistance of Pile 1, Pile 2, Pile 3, Pile 4 increases by 45.5%, 55.2%, 56.3%, 53.1% respectively compared with the previous pile jacking step. In contrast, for jacking depth 650mm, the pile tip resistance increases by 15.6%, 11.7%, 11.6%, 10.6% respectively, indicating that the critical depth is about 550mm (i.e., 11D). The responses of pile tip resistance are similar to that of jacking installation process conducted by Zhou et al. (2009).

According to the variation curves in Figure 6.7(c), it can be observed that the jacking resistance of multi-pile foundation is related to jacking sequence. In other words, the jacking resistance of the latter pile is larger than that of the former pile. The main reasons for this phenomenon are that vertical displacement of sand relative to pile is generated because of downward penetration of the former pile. Sand surrounding pile is compressed radially. As a result of this, the latter pile is subject to larger resistance owing to densification of sand.

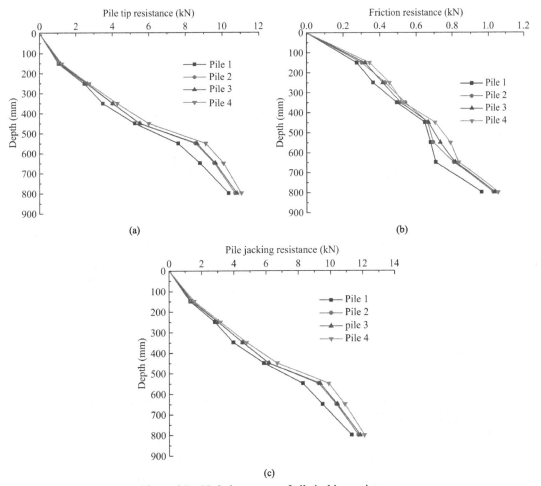

Figure 6.7 Variation curves of pile jacking resistance
(a) Pile tip resistance; (b) Frictional resistance; (c) Total pile jacking resistance

Axial force of pile

Figure 6.8 depicts the relationship between axial force of pile and penetration depth. It can be seen that axial force of pile increases gradually with the increase of penetration depth. This is due to the fact that, during jacking installation, the axial force is bound to follow the static equilibrium condition. Thus, with the increase of friction and pile tip resistance, the axial force of pile increases. At the end of jacking installation, the axial force in Pile 1 is 14.48kN, 14.20kN, 13.84kN, 13.35kN, 12.70kN, 12.02kN, 11.31kN, 10.35kN at 50mm, 150mm, 250mm, 350mm, 450mm, 550mm, 650mm, 750mm depth respectively. Same trend is displayed in other three piles. Therefore, it can be concluded that axial force of pile decreases gradually along the pile depth when penetration depth is constant. Furthermore, the decreasing amplitude is increasing. The main reasons can be attributed to the fact that the interface shear strength is reached due to continuous downward shear in shallow sand. For the deep sand, because pile tip resistance can offset part of jacking force, the frictional resistance has not fully mobilized.

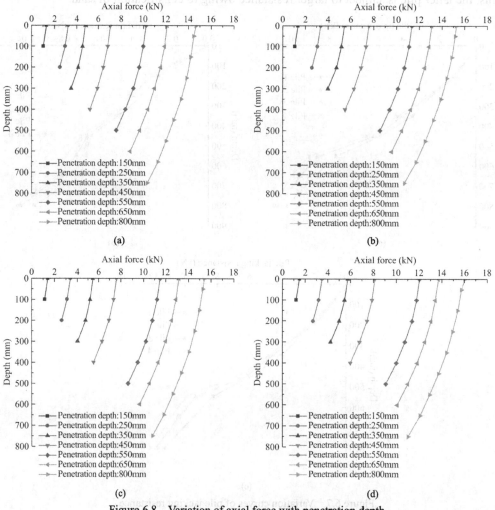

Figure 6.8 Variation of axial force with penetration depth
(a) Pile 1; (b) Pile 2; (c) Pile 3; (d) Pile 4

The axial force of four piles is compared when penetration depths are 450mm and 800mm respectively, as shown in Figure 6.9. The trend shows a good agreement with that of jacking resistance. At the same depth, the axial force of the latter pile is greater than that of the former pile. This is also attributed to the compaction effect. With the increase of depth, the axial force of the latter pile is gradually close to that of the former pile. Two reasons can be found to explain the above trends in this study. On one hand, shallow sand becomes denser due to the compaction effect, resulting in larger frictional resistance of the latter pile (note that the axial force is calculated from frictional resistance). On the other hand, the contribution of friction to balancing the jacking force is small for the bottom of pile. After pile penetrating to the critical depth, pile tip resistance gradually reaches its limit value. Hence, the axial force of four piles tends to be close.

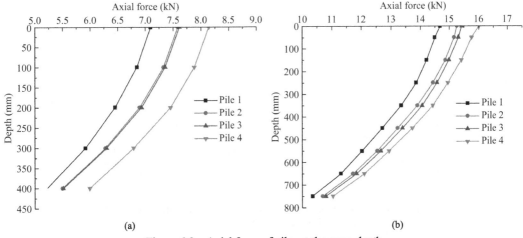

Figure 6.9 Axial force of piles at the same depth
(a) Penetration depth 450mm; (b) Penetration depth 800mm

Frictional resistance

The variation of unit frictional resistance with the increase of depth is displayed in Figure 6.10. It can be summarized that the unit frictional resistance of pile increases nonlinearly with the increase of penetration depth. In addition, the larger the penetration depth, the greater the maximum unit friction of pile. As pile is jacked downwards, pile-soil interaction is strengthened continuously due to the high stress level of deep sand. At the same depth, the larger the penetration depth, the smaller the unit frictional resistance. This can be explained as follows. In the course of pile penetration, sand particles in contact with pile are constantly rearranged. Large sand particles are ground, and the generated fine particles enter the pores of sand, resulting in the decrease of friction coefficient of pile-soil interface. The "friction fatigue effect" can also be interpreted by the results of unit frictional resistance.

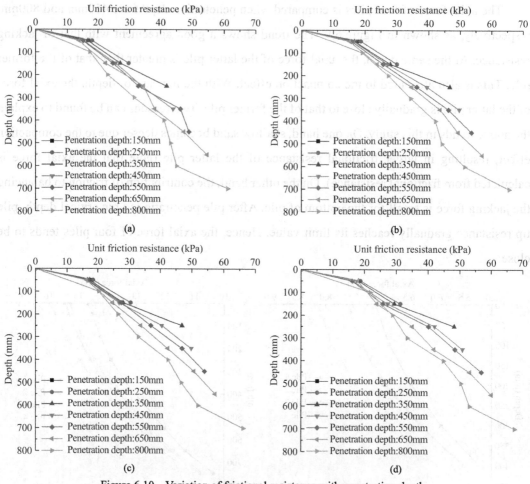

Figure 6.10 Variation of frictional resistance with penetration depth
(a) Pile 1; (b) Pile 2; (c) Pile 3; (d) Pile 4

6.1.2.2 Static loading responses

Bending moment of pile

The bending moment of piles versus depth is depicted in Figure 6.11. With the increase of static load, the bending moment of pile at different depths increases, and the maximum bending moment point moves downwards in a certain depth range. The maximum bending moment point of the back row of piles is about $4D$-$5D$ below the ground, while that of the front row of piles is about $3D$-$4D$. The depth of bending moment reverse point is about $11D$ for both front and back rows of piles. The overturning moment caused by static load is transferred to the piles of jacket foundation mainly in axial loading form. The front row of piles is pushed downwards, while the back row of piles is pulled upward. Due to the difference of axial force the stress state of sand around pile is altered, resulting in the different responses of sand resistance. As a result of this, there is discrepancy in bending moment response between the front and back rows of piles.

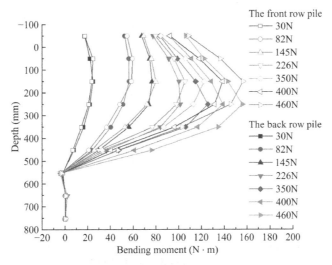

Figure 6.11 The bending moment variation of front and back rows of piles

Displacement of pile

Figure 6.12 illustrates the horizontal displacement versus the depth of front and back rows of piles. The larger the static load, the greater the displacement value. The horizontal displacement rapidly decreases with the increase of depth and mainly occurs within $9D$ below the ground. The displacement values of front and back rows of piles are almost the same. It is indicated that the distance between piles is constant and jacket foundation only tilts at a certain angle. Through the measured bending moment, it can be deduced that the variation of soil resistance around pile with depth is close to hyperbolic form. The $P\text{-}y$ curves of front and back rows of piles at depth of 50mm and 150mm are displayed in Figure 6.13. Within $3D$ depth, the horizontal resistance of sand around the front row of pile is larger than that around the back row of pile. In addition, the greater the depth, the faster the resistance growth in early stage, and the slower the resistance growth in later stage.

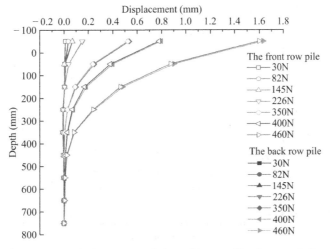

Figure 6.12 Displacement variation of front and back rows of piles

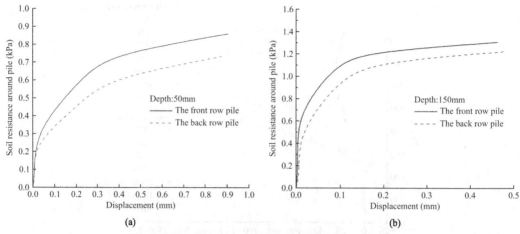

Figure 6.13 *P-y* curves of front and back rows of piles
(a) Depth: 50mm; (b) Depth: 150mm

6.1.2.3 Cyclic loading responses

Horizontal displacement at tower top

The evolution of maximum horizontal displacement at tower top against cycle number are presented in Figure 6.14. In light of the displacement growth rate for T1 to T5, the displacement curve undergoes two stages during cyclic loading: rapid growth stage (about the first 10000 cycles) and slow growth stage. The displacement values at 10000 cycles account for 76.0%, 79.7%, 87.0%, 88.7% and 87.1% of the termination displacement respectively. Then, the displacement growth rate slows down and finally tends to be stable with the increase of cycle number. This finding is similar to the displacement development mode in centrifuge model test (Rosquoet et al., 2007). Moreover, the larger the load frequency or cyclic load ratio, the bigger the horizontal displacement.

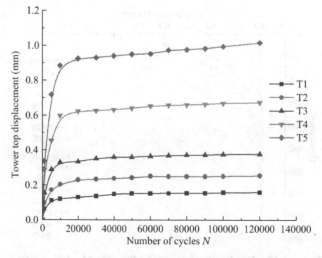

Figure 6.14 Horizontal displacement variation at tower top

Frictional resistance of pile

The service failure of OWT jacket foundation is controlled by the back row of pile (tension piles). The frictional resistance of the back row of pile versus depth is described in Figure 6.15. Under different cycles, the variation trends of unit frictional resistance with depth are almost the same. Frictional resistance increases first and then decreases along depth, showing a skewed V shape. The maximum value point of unit friction occurs about 8D below the ground. The main reason for the variation trend of unit friction is attributed to the change of sand relative density around pile in jacking installation. For shallow sand, vertical heave is the main deformation form. The porosity of shallow sand increases, resulting in the decrease of particle friction coefficient. Therefore, the unit frictional resistance is relatively small. With the increase of depth, the main deformation of middle sand transforms to radial compression. The porosity of sand decreases, and the particle friction coefficient increases. Hence the unit frictional resistance is relatively large. Nevertheless, the influence of cyclic loading on frictional resistance becomes insignificant for the depth below 8D.

Figure 6.16 depicts the relationship of unit frictional resistance versus cycle number. For depth less than 8D, the frictional resistance gradually decreases first with the increase of cycle number. Then it tends to be stable after 10000 cycles. When depth is greater than 10D, the frictional resistance first increases slowly with the increase of cycle number. After about 5000 cycles it decreases gradually. Finally, the frictional resistance hardly varies after the same 10000 cycles. Shallow sand is strongly disturbed by the cyclic loading, resulting in radial displacement and particle breakage of sand around pile. With the continuous cyclic loading, the effective stress of pile-soil interaction decreases, and finally the frictional resistance keeps stable. In contrast, for the sand with depth below 10D, there is little displacement and breakage in the early stage of cyclic loading. Considering the high stress level of sand, the frictional resistance has a growth stage until sand particles are settled.

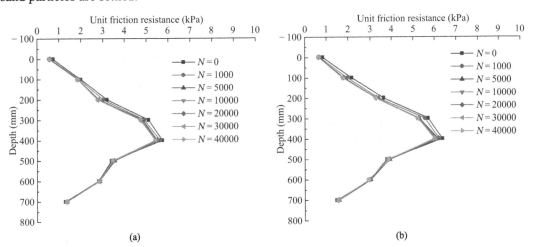

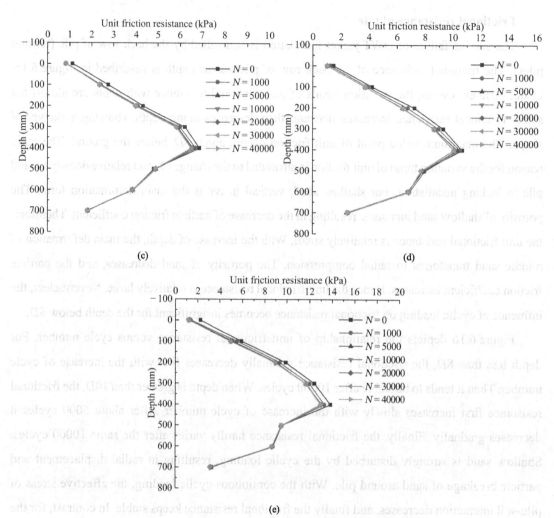

Figure 6.15 Frictional resistance of back row of pile versus depth
(a) T1; (b) T2; (c) T3; (d) T4; (e) T5

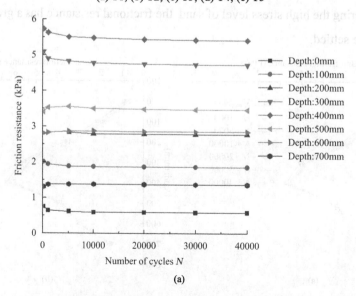

(a)

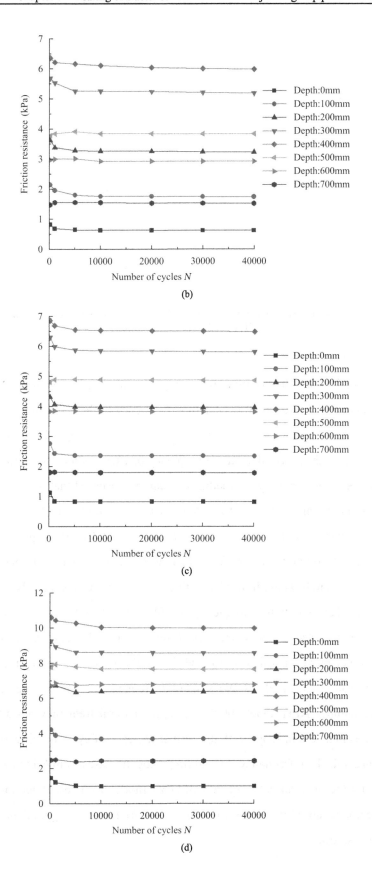

(b)

(c)

(d)

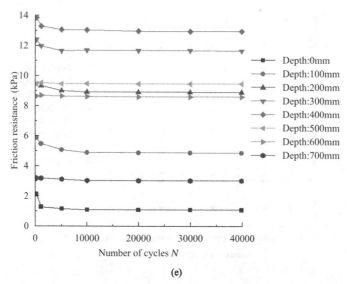

(e)

Figure 6.16 Frictional resistance of back row of pile versus cycle number
(a) T1; (b) T2; (c) T3; (d) T4; (e) T5

Natural frequency of OWT

In this section, the normalized natural frequency calculated by Fast Fourier Transform (FFT) is employed to characterize the variation of natural frequency of the OWT model with increasing cycle number. Figure 6.17 depicts the experiment results for the changes in natural frequency, in which $f_{initial}$ refers to the first-order natural frequency measured after pile jacking and f refers to that after a certain number of cyclic loadings. It can be observed that the normalized natural frequency varies in the range from 0.95 to 1.05 in tests T1 to T3. In test T1 with small excitation frequency (load frequency is 6Hz), the natural frequency of OWT model experiences an obvious increase before about 10000 cycles and then decreases slightly. Finally, the natural frequency keeps oscillating in a small range. In test T2 with excitation frequency slightly less than $f_{initial}$ (load frequency is 8Hz), the natural frequency of OWT model first decreases and then sharply increases with the increase of cycle number. After a minor descent, the natural frequency fluctuates in a larger range than that of T1 and T3. While in test T3 (load frequency is 12Hz slightly greater than $f_{initial}$), the OWT's natural frequency obviously decreases with the increasing cycle number. After about 10000 cycles, the natural frequency has a tiny increment and then oscillates in a certain scope with the further increase of cycle number. In addition, it is noted that the larger the load frequency is (load frequency increases from 6Hz to 12Hz in tests T1 to T3), the smaller the final natural frequency of OWT model becomes. When load frequency is smaller than the initial natural frequency, the final natural frequency is bigger than its initial value. The converse is also true.

The relationship between cyclic load ratio and natural frequency of OWT model is shown in Figure 6.18. It is indicated that the normalized natural frequency varies in the range from 0.908 to 1 in tests T3 to T5. For tests T3 to T5 with different cyclic load ratios, in which the load frequency is set as 12Hz, the natural frequency of OWT model decreases in varying degrees before about 10000 cycles. It can be easily observed that the bigger the cyclic load ratio (load magnitude), the faster the decrease speed. Furthermore, with the further increase of cycle number, the natural frequency slightly increases (except in test T4) and then fluctuates in a certain range. The fluctuation range increases with the increase of cyclic load ratio. Moreover, the bigger the cyclic load ratio is (load magnitude increases from 38N to 114N in tests T3 to T5), the smaller the final natural frequency of OWT model becomes. Regardless of the value of cyclic load ratio in this study, the final natural frequency is always smaller than its initial value.

In the process of cyclic loading, the sand around pile is disturbed in different degrees when loading conditions such as load frequency and load magnitude change. The main reason for the change of natural frequency is reflected in the variation of foundation stiffness. With the continuous action of two-way cyclic excitation, sand volume strain gradually accumulates and the sand around pile becomes denser. The constraint enhancement of surrounding sand has a significant influence on the foundation stiffness of OWT. Meanwhile, sand subsidence around the embedded piles is observed in cyclic loading tests. Under extreme load conditions, the foundation inclination caused by cyclic loads weakens the constraints of surrounding sand. Therefore, the decrease of natural frequency can be interpreted by migration mechanism of sand particles.

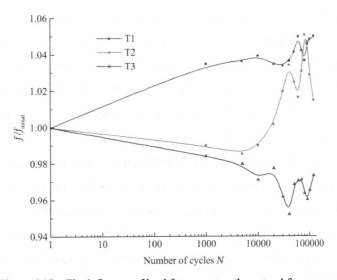

Figure 6.17   The influence of load frequency on the natural frequency

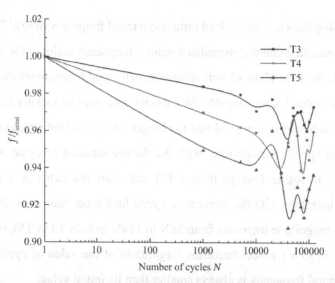

Figure 6.18 The influence of cyclic load ratio on the natural frequency

Damping ratio of OWT

System damping is also one of the important parameters to describe the dynamic characteristics of structure. The ratio of the system damping ratio of OWT model after cyclic loading to its initial system damping ratio, i.e., $\zeta/\zeta_{initial}$, is used to depict the damping change of OWT model. Figure 6.19 presents the changes of system damping ratio calculated by logarithmic attenuation method with increasing cycle number. The ζ and $\zeta_{initial}$ refer to system damping ratio after cyclic loading and pile jacking respectively. The influence of load frequency on system damping ratio is shown in Figure 6.19(a). It is indicated that the normalized system damping ratio varies in the range from 0.4 to 1.02 in tests T1 to T3. For tests T1 and T2, the system damping ratio continuously decreases and finally fluctuates in a certain range. While in test T3, the system damping ratio experiences an extremely slow growth before about 1000 cycles. Then system damping ratio decreases and finally starts oscillating. Hence, it can be inferred that when load frequency is bigger than natural frequency of OWT, the system damping ratio will experience a growth stage. Further, the larger the load frequency, the greater the increase magnitude of system damping ratio. The converse is also true. In addition, the system damping ratio in three tests occurs a sharp decline at different cycles respectively.

Figure 6.19(b) displays the variation of system damping ratio with respect to cyclic load ratio. It can be easily observed that there is a growth phase of system damping ratio in early stage of cyclic loading before 10000 cycles. The greater the cyclic load ratio, the greater the peak value of system damping ratio and the greater the cycle number to reach the peak value of damping ratio. Large cyclic load ratio corresponds to a big growth rate of system damping ratio in early stage. Then, the system damping ratio undergoes a significant decline and finally remains irregular oscillation.

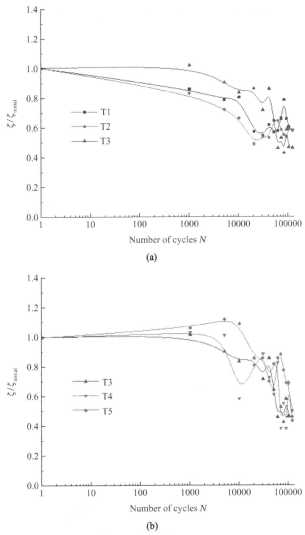

Figure 6.19 The variation of system damping ratio
(a) The influence of load frequency; (b) The influence of cyclic load ratio

Comparison of natural frequency and system damping ratio

In order to investigate the relationship between natural frequency and system damping ratio of OWT model, the experiment results of cyclic loading are plotted together, as shown in Figure 6.20. It can be summarized that the evolution trend of system damping ratio and natural frequency is roughly opposite. When natural frequency increases or remains stable, system damping ratio decreases. While natural frequency decreases, system damping ratio increases. With the densification of sand, the system damping ratio generally presents a descent trend in the five tests. Further, with the increase of cycle number, the final system damping ratio is smaller than its initial value. Compared with the natural frequency, the system damping ratio has a bigger change range, with a maximum reduction of 64%. The decrease of damping ratio is one order of magnitude larger than the increase of natural frequency. Hence, in addition to making the natural frequency of OWT meet the requirements, the variation of system damping ratio should also be paid enough attention.

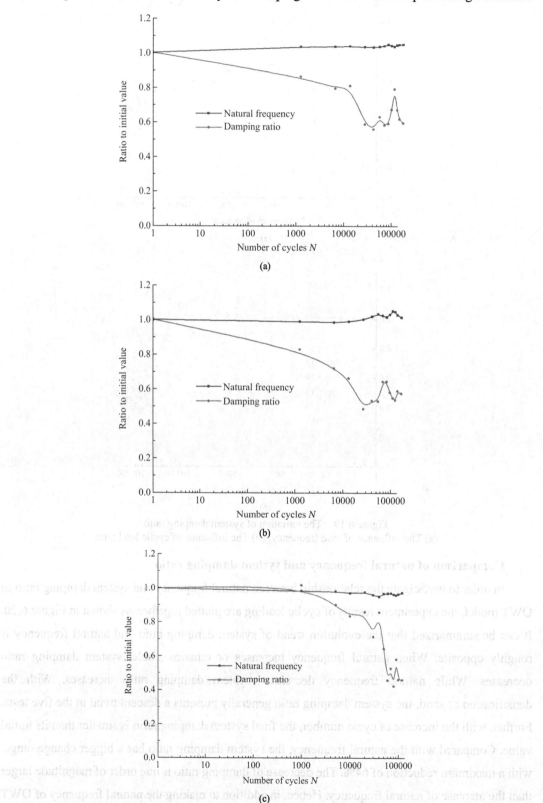

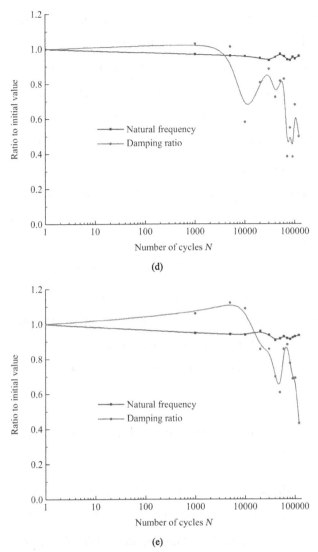

Figure 6.20　Comparison of natural frequency and system damping ratio
(a) T1; (b) T2; (c)T3; (d)T4; (e)T5

6.1.3　Summary

In this chapter, a series of large-scale indoor model tests were conducted to investigate the mechanics of pile jacking installation and the lateral loading performance of jacket supported OWT in dense sand. The effects of load frequency and cyclic load ratio on the dynamic responses of jacket foundation were discussed. Based on the experiment results, the following conclusions can be drawn:

(1) For pile jacking installation, the linear growth of pile tip resistance can be divided into two stages: rapid growth stage and slow growth stage, and its critical depth is about $11D$. The jacking resistance of multi-pile foundation is related to the jacking sequence. The jacking

resistance of the latter pile is larger than that of the former pile. With the increase of penetration depth, the axial force increases while the unit frictional resistance decrease.

(2) Under horizontal static load, the maximum bending moment point moves downwards in a certain depth range. The maximum bending moment points of the back row of piles and the front row of piles are about $4D$-$5D$ and $3D$-$4D$ below the ground respectively. The depth of bending moment reverse point is about $11D$ for the front and back rows of piles. The horizontal displacement mainly occurs within $9D$ below the ground.

(3) Under horizontal cyclic loads, the displacement at tower top undergoes two stages. The larger the load frequency or cyclic load ratio, the bigger the horizontal displacement. The frictional resistance of the back row of piles increases first and then decreases along depth, showing a skewed V shape. The maximum value point of unit friction occurs about $8D$ below the ground. The response of friction above $8D$ and below $10D$ is different.

(4) The natural frequency varies with different load frequencies and cyclic load ratios. The evolution trend of system damping ratio and natural frequency is roughly opposite. Furthermore, the change of damping ratio should also be paid enough attention.

6.2 Dynamic response of open-ended pipe pile under vertical cyclic loading in sand and clay

Lots of research for the bearing characteristics of piled foundation have been carried out. Most of these studies focused on the bearing characteristics of piles under static load (Poulos, 1989; Janbu, 1976; Ottaviani, 1975; Burland, 1973; Chandler, 1968; Skempton, 1959). As pointed out by Poulos (1989), the shear weakening effect of the pile-soil interface under cyclic loading is more obvious than that under static loading. Until now, numerical simulation methods are majorly utilized to study the bearing characteristics of piles under dynamic load (Liu et al., 2019; Mohammad et al., 2011; Arroyo et al., 2011; Chaudhary, 2007; Trochanis et al., 1991; Pressley et al., 1986; Muqtadir et al., 1986; Huang et al., 2015). However, inappropriate selections of parameters in numerical simulations will cause errors in the prediction. Therefore, an experimental approach is a direct and fundamental way to reveal the characteristics of the pile foundation under cyclic loading, and it can be used to validate the theoretical models. In the past, many researchers have carried out vertical cyclic load tests on pile foundations in different foundation soils, including calcareous sand, sandy soil, clay and other types of foundation soils (Al-Douri et al., 1995; Al-Douri et al., 1992; Kraft et al., 1981; Seed et al., 1955; Tsuha et al., 2012). However, most previous studies were small-scale tests in one type of soil. In their study, the size effects of experiments may affect the results, and they are lacked a comparative analysis of pile bearing characteristics between different soils.

On the other hand, double-walled pipe piles were used to examine the internal and external frictional resistance of an open-ended pipe pile. Later, Choi and O'Neil (1997) investigated the resistance of soil plugs in open-ended pipe piles driven into saturated sand under seismic load. Paik et al. (2003) studied that the dynamic responses of soil plugs to open-ended piles and the bearing capacity of pipe piles. The cumulative hammer blow count for the open-ended pile was 16% lower than for the closed-ended pile. After that, the dynamic load-bearing characteristics of open-ended tubular piles were studied (Igoe et al., 2010; Iskander, 2010; Paik et al., 2003; Jeong et al., 2015). However, there are few studies on the bearing behavior of open-ended pipe piles under different loading frequencies in different soil layers.

A lot of studies for the bearing characteristics of open-ended pipe piles have been carried in the past, but effects of soil layers and load frequencies have not been considered. In order to better reveal the bearing mechanism of open-ended pipe piles. In this study, a series of large-scale model tests were conducted to explore the effects of different soil layers and load frequencies on the bearing characteristics of open-ended pipe piles. The outcomes of this study can provide a reliable basis for actual engineering design.

6.2.1 Materials and Methods

6.2.1.1 Large-scale model box

This experiment was conducted in a large-scale model box, with the dimension 3000mm × 3000mm × 2000mm (length × width × height). The control device of the model box was composed of a loading system, an oil pressure control system, a data collection system, and a drainage system, as shown in Figure 6.21.

In this study, the model pile was designed as a double-walled open-ended pipe pile to measure the internal and external friction of the open-ended pipe pile. The model pile is connected by two aluminum alloy round pipes through pile shoes. The Poisson's ratio of the aluminum alloy pipe is 0.3, and the elastic modulus is 72GPa. Figures 6.22 and 6.23 depict the dimensions of the model pile. The model pile is 100mm in length, 140mm in diameter, and 13mm in thickness. The optical fiber grating miniature strain sensor is pasted in a reserved groove on the outer wall of the outer tube and sealed with epoxy resin. The outer wall of the inner tube is a closed annular space, and the sensors can be directly attached to the outer wall of the inner tube to ensure the safety of the sensors.

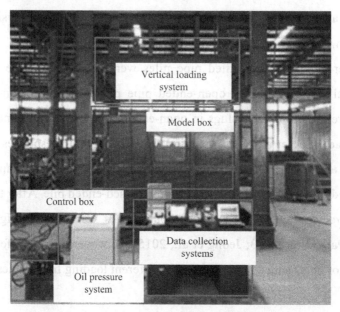

Figure 6.21 Large-scale model test control system

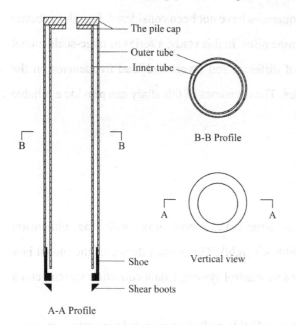

Figure 6.22 The profile of double-walled pile

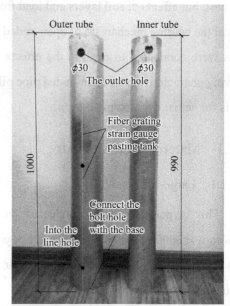

Figure 6.23 The inner and outer tubes of double-walled open-ended pile

6.2.1.2 Material for tests

The sand soil used in this study was naturally air-dried sea sand of Qingdao, and the clay was collected from the coastal region of Qingdao city, Shandong province, China. The basic geotechnical characteristics of the soil samples were shown in Tables 6.4 and 6.5.

Physical characterization of testing sand soil Table 6.4

Physical parameter	Value
Density (kg/m³)	2650
Median particle size d_{50} (mm)	0.72
Relative compactness D_r	0.73
Internal friction angle φ (°)	42.8
Maximum void ratio e_{max}	0.52
Minimum void ratio e_{min}	0.30
Dry density ρ_d (kg/m³)	1950

Physical characterization of testing clay Table 6.5

Physical parameter	Value
Liquid limit (%)	42.7
Plastic limit (%)	27.5
Plasticity index	15.2
Specific gravity	2.53
Maximum dry density (kg/m³)	1780
Optimum moisture content (%)	14.7

6.2.2 Experimental Procedure

(1) Fill the soil sample into the model box by layer each with a height of 200mm. Each layer of soil sample shall be tamped manually twice and by machine once until reaching the required test elevation. Each layer of sand soil sample was allowed to stand for at least 12h, and the clay soil sample was allowed to stand for at least 24h.

(2) Before penetration the pile, first move the hydraulic jack on the beam to the designated pile position through the electronic control system, raise the jack to a certain height, place the model pile at the designated position, and use a level to detect whether the pipe pile was vertical to prevent eccentric compression. After the pipe pile was upright, the hydraulic jack was controlled to descend slowly and uniformly to achieve uninterrupted penetration under static pressure until the design elevation. The pile penetration speed was about 300mm/min.

(3) After the pile penetration was completed, the soil sample was allowed to stand for at least 20 days before the loading test. Ensure that the center of the loading device coincides with the center of the pile

(4) Static load tests were carried out in a stepwise loading manner. The load of the first step was 10kN, and the load of each step was increased by 5kN. When the settlement at the top of the

model pile was greater than twice of the previous step or the total settlement at the pile top was greater than 40mm, the model pile was considered to be damaged (JGJ 94—2008). The dynamic load was applied to the model pile through the servo loading device. Table 6.6 presents the test scheme for the dynamic loading test.

In order to accurately simulate the dynamic characteristics of the model structure, the actual load frequency and natural vibration frequency on site are equivalent to the model load frequency and natural vibration frequency according to a specific relationship (Bhattacharya et al., 2011), as shown in Equation (6.1).

$$\left(\frac{f}{f_n}\right)_{model} = \left(\frac{f}{f_n}\right)_{prototype} \tag{6.1}$$

The 1P frequency in the prototype of the wind turbine is 0.082-0.216Hz. The natural vibration frequency of the wind turbine prototype is about 0.289Hz. The measured natural vibration frequency of the open-ended and closed pile foundation structure is about 9Hz, and the 1P frequency of the test model structure calculated corresponds to 2.55-6.68Hz (1P). Finally, the Dynamic load amplitude of 3Hz and 5Hz were selected for the tests.

Dynamic load test scheme of piles Table 6.6

Test number	P1	P2	P3	P4
Dynamic load amplitude (kN)	3	3	3	3
Median dynamic load (kN)	1.5	1.5	1.5	1.5
Waveform	Sine wave	Sine wave	Sine wave	Sine wave
Frequency (Hz)	3	5	3	5
Cycle number	1000	1000	1000	1000
Soil sample	Sand	Sand	Clay	Clay

6.2.3 Results and Discussions

6.2.3.1 Static Load Test

Static load tests were conducted to determine the bearing capacity of open-ended pipe piles. These tests will provide a theoretical basis for determining the cyclic load ratio of the vertical cyclic load applied to the pile top. ζ_b was the degree of cyclic load application proposed by Byrne et al. (2010), which could be expressed as:

$$\zeta_b = \frac{H_{max}}{H_u} \tag{6.2}$$

where H_u denotes the vertical ultimate bearing capacity, and H_{max} denotes the dynamic load amplitude.

In the experiments, the pile is considered to be damaged under one of the following conditions: (1) when the settlement at the top of the pile was greater than twice that of the previous step or (2) the total settlement at the pile top was greater than 40mm. The experimental results show that the pile displacement increases with increasing static load. With the same vertical load at the top of the pile, the settlement of the pile in silty clay is greater than the settlement of the pile in sand. The ultimate bearing capacity of an open-ended pipe pile in clay is $Q_{us} = 6$kN. When the ultimate load is reached, the cumulative settlement at the top of the pile is $S_{us} = 4.88$mm. The ultimate bearing capacity of open-ended pipe piles in sand is $Q_{us} = 30$kN. Upon reaching the ultimate load, the cumulative settlement of the pile top $S_{us} = 5.35$mm. It could be concluded that the ultimate bearing capacity of open-ended pipe piles in sand is much larger than that in clay. In this study, the cyclic load ratios in sand and clay were measured to be 0.1 and 0.5 respectively. The dynamic load amplitude of 3kN was selected in the tests.

6.2.3.2 Cyclic Dynamic Load Tests

Under the action of cyclic load, the outer frictional resistance of the pile-soil contact surface will be weakened under the action of cyclic load, and the inner frictional resistance of the pile-soil plug contact surface will be weakened. Dynamic load tests were conducted to study the weakening mechanism of pile-soil contact surface.

Cumulative displacement settlement

Figure 6.24 shows the cumulative displacement of pile top under cyclic dynamic load. It could be seen that the cumulative settlement at the top of the model pile increased rapidly. When the number of cycles reached 200, the cumulative settlement at the pile top remained basically stable. For the same soil sample, the speed of settlement at the top of the model pile at the 5Hz vibration frequency is larger at the initial stage, and the final cumulative displacement is larger than the trend obtained for 3Hz vibration frequency. It shows that the vibration frequency has a great impact on settlement of the pile top. Under the same vibration frequency, the settlement of pile top in clay increases at an early stage, and the final cumulative displacement is larger than that in the sand.

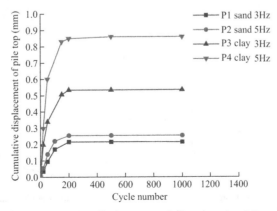

Figure 6.24 Cumulative displacement of pile top under different cycles

Axial force of pile

Figure 6.25 illustrates the distribution of the variation of the axial force of the outer and inner pipe of a double-walled pipe pile versus the soil depth in sand under cyclic dynamic loading. It can be seen from the figure that, under the dynamic load in sand, the axial force of the outer pipe pile gradually decreases with increase in depth, and the rate of decrement is fast first and then slows down. When the pile is subjected to a vertical cyclic load, the vertical load is transmitted along the axial force of the pile. The load needs to continuously overcome the lateral frictional resistance during the load transfer process, so the axial force of the pile body gradually decreases. As shown in Figures 6.25(a) and (b), with the number of cycles increased, the axial force of the outer tube of the P1 pile increased by 1.4%, 1.8%, 1.9%, 1.6%, 1.7%, 1.1% from top to bottom, the axial force of the outer tube of the P2 pile increased by 1.57%, 1.95%, 2.3%, 1.6%, 1.77%, 1.2% from top to bottom. It shows that, for sandy soil, the axial force of the outer pipe pile increases gradually with the increase in the number of cycles. Under the same number of cycles, the larger the vibration frequency, the higher the axial force at the same position of the outer pile increases.

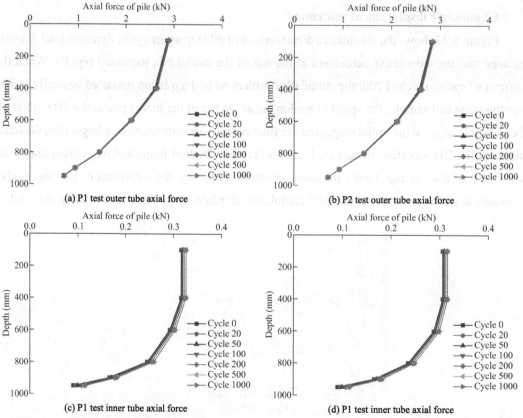

Figure 6.25 The distribution of the variation of axial force of pile body in sand

From Figures 6.25(c) and (d), with increasing number of cycles, the axial force of the inner tube of the P1 pile increased by 1.4%, 1.8%, 1.9%, 1.6%, 1.7%, 1.1% from top to bottom, while the axial force of the inner tube of the P2 pile increased by 1.57%, 1.95%, 2.3%, 1.6%, 1.77%, 1.2% from top to bottom. It can be seen from the figures that under the action of cyclic load, the axial force of the inner pipe pile gradually increases with the increase in the number of cycles. Under the same number of cycles, the larger the vibration frequency, the more the axial force at the same position of the inner pile increases. The increase in the axial force of the inner tube is smaller than that of the outer tube.

Variations of the axial force of the inner and outer pipe shafts of double-walled pipe piles in clay with depth under cyclic dynamic loading are plotted in Figure 6.26. Under cyclic loading, the axial force of the pile shows a non-linear distribution and gradually decreases from top to bottom. As the number of vibrations increases, the axial force of the pile increases slightly. Under the cyclic dynamic load of 3Hz, the axial force of the P3 outer pipe pile body increased from top to bottom by 2.9%, 3.5%, 5.0%, 6.6%, 7%, 6.2%. The axial force of the inner pipe pile shaft increased by 0.41%, 0.41%, 0.42%, 0.44%, 0.483%, 0.55%. Under the cyclic dynamic load of 5Hz, the axial force of the P4 outer pipe pile body increased from top to bottom by 3%, 3.9%, 5.3%, 7%, 7.5%, 6.5%, The axial force of the inner pipe pile shaft increased by 0.43%, 0.43%, 0.445%, 0.455%, 0.49%, 0.575%. The axial force of the outer tube decreases rapidly at a depth of 600mm due to measurement errors. Under the dynamic load of 5Hz, the axial force increase of the inner and outer piles is greater than 3Hz. This inference indicates that under the same number of cycles, the higher the dynamic load frequency, the greater the axial force of the pile. Comparative analyses of Figure 6.25 and Figure 6.26 show that, with the same dynamic load frequency and number of cycles, the axial force of the outer tube in the clay decreases faster than that in the sand. The main reason is that the lateral frictional resistance needs to be overcome when the axial force is transmitted from the top to the bottom. Under the same number of cycles, the cumulative displacement and settlement of the pile in the clay is larger, and the lateral frictional resistance of the pile in the clay is higher.

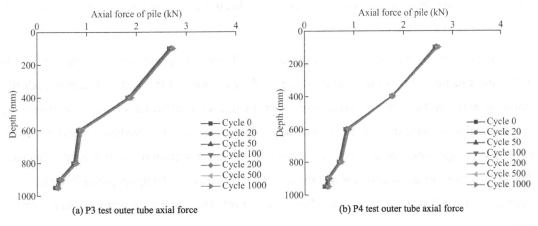

(a) P3 test outer tube axial force   (b) P4 test outer tube axial force

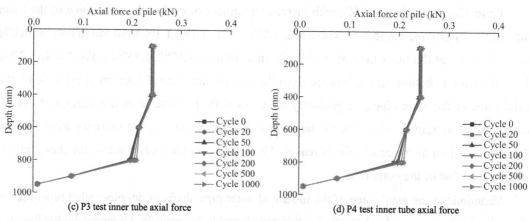

(c) P3 test inner tube axial force (d) P4 test inner tube axial force

Figure 6.26 Variation of axial force of pile in clay

Shaft friction of pile

Figure 6.27 shows the variations in unit external friction of the pile body in sand with depth under different dynamic load cycles and variation in the unit external friction of the pile body at different depth sections of the pile with the number of cycles. Here the depth is defined as the distance from the section to the top of the pile. From Figures 6.27(a) and (c), under different cycles, the unit external friction of an open-ended pipe pile in sand is skewed V shaped with increasing depth. The unit external friction of the pile increases firstly and then decreases with increasing depth. The main attributions are that, during the application of dynamic load, the shaft friction of the pile body gradually exerts from the top to the bottom. On the upper part of the pile, the lateral soil pressure of the pile gradually increases with increasing depth, which leads to the increase of the unit lateral friction of the pile. the shaft frictional resistance is exerted to a large level. On the other hand, due to the axial compression of the pile body, the settlement of the lower pile body is smaller than that of the top of the pile, and the relative movement of the pile body and the soil is smaller. The side frictional resistance is not fully mobilized, so the unit external frictional resistance of the pile body gradually decreases, and this trend is similar to the research results in reference (Liu et al., 2004).

From Figures 6.27(b) and (d), as the dynamic cyclic load progresses, the external unit friction at the lower depth section of the pile decreases at the beginning of the cyclic load, and gradually stabilizes as the cyclic load progresses. The external unit friction at the middle depth section of the pile has a clear upward trend at the beginning of the cyclic load, and gradually stabilizes as the cyclic load progresses. The unit external friction of the lower section of the pile hardly changes. The main reason for this analysis is that the initial shaft friction of the upper part of the pile at the beginning of the dynamic load first reaches a large level. The soil particles at the shear interface are rearranged, shredded and refined with the increasing number of cycles. Consequently, the lateral pressure at the pile-soil interface is gradually reduced, and the friction coefficient between the pile and soil decreases. Therefore, the frictional resistance on the side of the pile gradually decreases. It is mentioned (Airey et al., 1992) that the side frictional resistance of pile will deteriorate under cyclic shear. As the dynamic load progresses, the shaft friction in the upper part of the pile gradually reaches its maximum value and are transmitted downwards. The unit external friction in the middle part of the pile gradually increased. The unit side frictional resistance of the lower part of the pile body has not been greatly affected.

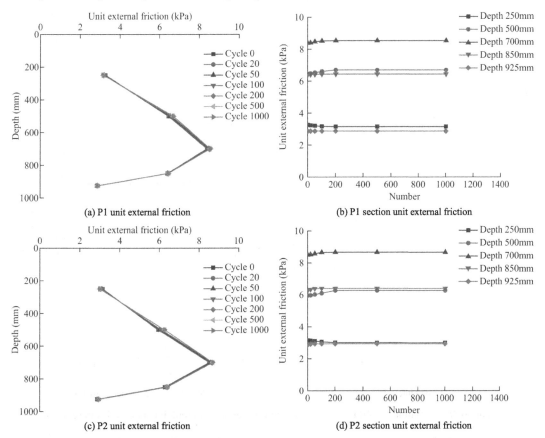

Figure 6.27 Variation of unit external friction of piles under different dynamic cycles in sand soil

Figure 6.28 shows the variations in unit internal friction of the pile body in sand with depth under different dynamic load cycles and variation in the unit friction of inside of the pile body at different depth sections of the pile with the number of cycles. As shown in Figures 6.28(a) and (c), under cyclic loading, as the depth increases, the unit side frictional resistance increases. This is mainly due to the dense soil plug inside the pile, which can provide great unit side frictional resistance.

It can be seen from Figures 6.28(b) and (d) that, the unit friction of the inner side of the deeper section of the pile in sand has a significant reducing trend at the beginning of the cyclic load, and gradually stabilizes as the cyclic load progresses. While the inner frictional resistance at the upper and middle sections of the pile does not change. The main attributions are that, under the cyclic load of the pile, the lower soil particles begin to re-arrange, making the lower soil inside the pile loose, and the inner frictional resistance gradually decreases until it is stable. The internal friction of the pile body reduces from the bottom to the top because that the plugging effect causes the soil along the lower part of the pile jammed and densified, while the soil along the upper and middle sections of the pile is still loose.

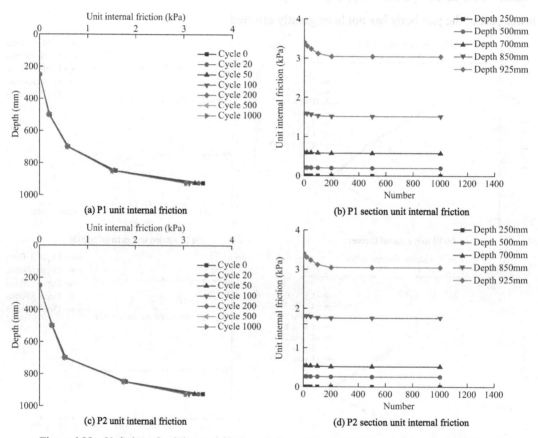

Figure 6.28 Variation of unit internal friction of piles under different dynamic cycles in sand soil

Figure 6.29 shows the variations of unit external friction of piles under different dynamic cycles in clay with increase in depth. Figures 6.29(a) and (c) illustrate the distributions of variation in unit external friction of the pile body in clay with depth under different dynamic load cycles. Figures 6.29(b) and (d) display the distributions of variation in the unit external friction of the pile body at different depth sections with the number of cycles. As shown in Figures 6.29(a) and (c), we know that, as the depth increases, the shaft friction of the pile body increases first and then decreases, and the trend was followed once again. Under different cycles in clay, the unit external friction of an open-ended pipe pile is skewed "W" shaped with increasing depth. This is a result of the combined effect of the increase in soil pressure around the pile from the top to the bottom and the formation of the water film on the upper pile-soil interface with the cycle. Firstly, the unit external friction of pipe piles in clay increases, the main reason is that the soil pressure increases with the depth. Then the decreasing trend is due to the fact that the side friction of the lower part of the pile has not yet played a role and the upper pile-soil interface forms a water film under the action of dynamic cycling, which plays a better lubrication role. The reason for the shaft frictional resistance increases again is that the water film at the lower part of the pile has not been formed, and the coefficient of friction is large.

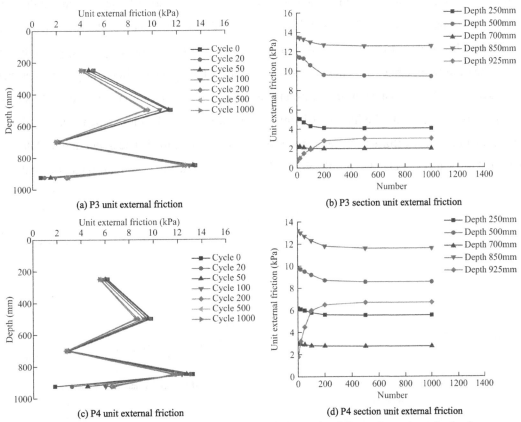

Figure 6.29 Variation of unit external friction of piles under different dynamic cycles in clay

Figure 6.30 shows the variations in unit internal friction of the pile body in clay with depth under different dynamic load cycles and variations in the unit internal friction of the pile body at different depth sections with the number of cycles. From Figures 6.30(a) and (c), it is seen that, under cyclic loading, the variation of the unit internal frictional along the depth of an open-ended pipe pile is similar to that in sand. From Figures 6.29(b) and (d), with the increase of the number of vibrations, the shaft frictional resistance of the upper and middle sides of the outer tube gradually decreases and then stabilizes, while the frictional resistance of the lower side of the outer tube gradually increases and then stabilizes.

The main reason for this analysis is that the frictional resistance of the upper and middle sides of the pile at the beginning of the dynamic load is first exerted to a large level. During the dynamic loading, as the number of cycles increases, a water film is formed at the pile-soil interface, which leads to a reduction in the friction coefficient and a reduction in frictional resistance. On the other hand, the frictional resistance of the lower side of the pile is small at the beginning of the dynamic load, and gradually mobilized with the application of the dynamic load.

When the dynamic load frequency is 5Hz, the range of the external frictional resistance is significantly larger than 3Hz, which indicates that the larger the load frequency, the higher the external frictional resistance. This result is consistent with the conclusion of the reference (Qu et al., 2020). The external friction of the middle side of the pile in the clay plays a bigger role than that in the sandy soil. Under the same load, the settlement of the pipe pile in the clay is greater than in the sand, and the external frictional resistance in the clay plays more significant roles. As shown in Figures 6.30(b) and (d), the internal friction of the lower part of the inner tube in clay gradually decreases and then stabilizes with the application of the dynamic load. The upper and middle side frictional resistances in clay have little changed, and this conclusion is similar to that of sand.

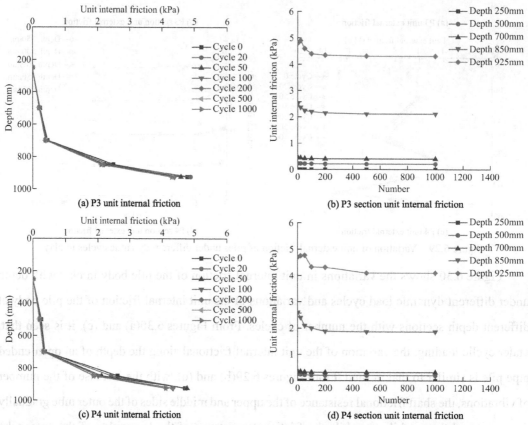

Figure 6.30   Variation of unit internal friction of piles under different dynamic cycles in clay

6.2.4 Summary

In this section, the vertical bearing characteristics of piles under static and dynamic loading conditions are examined through large-scale laboratory experiments. Based on experimental results, the main conclusions are as follows:

(1) The cumulative settlement of the pile top gradually increases with increasing static load. Under the same load, the displacement of the pile top in the clay is greater than the displacement in the sand. The ultimate bearing capacity of the open-ended pipe pile in the sand is much greater than that in clay.

(2) Under cyclic dynamic loading, the cumulative settlement of the pile top increases rapidly and then stabilizes. The cumulative settlement displacement of the pile top in the clay is greater than that in the sand. In the same soil layer, the greater the dynamic load frequency, the greater the cumulative settlement of the pile top.

(3) Under the action of dynamic load, the axial force of the pile gradually decreases with increasing depth. As the number of dynamic load cycles increases, the axial force of the pile body at the same location gradually increases, and the larger the vibration frequency, the greater the axial force of pile. The increase in axial force of the inner tube is smaller than that of the outer tube. In the same case, the axial force of the outer tube in the clay decreases faster than in the sand.

(4) Under dynamic loading, the distribution of the unit external frictional forces of pipe piles along depth in sand and clay shows skewed V shape and W shape. In the sand, the unit external friction of the upper part of pile decreases first and then stabilizes. The unit friction of the outer side of the middle part increases first and then stabilizes. The unit friction of the lower part changes little. In the clay, the unit external friction of the upper and middle piles in the clay decreases first and then stabilizes. The unit external friction of the lower part of the pile first increases and then stabilizes. The larger the load frequency, the more obvious the change in the unit external friction. In the sand and clay, the unit internal frictional resistance of the pile shows the same trend.

References

Airey D W, Al-Douri R H, Poulos H G. Estimation of pile friction degradation from shearbox tests. Geotechnical Testing Journal, 1992, 15(4): 388-392.

Al-Douri R H, Poulos H G. Predicted and observed cyclic performance of piles in calcareous sand. Journal of Geotechnical Engineering, 1995, 121(1): 1-16.

Al-Douri R H, Poulos H G. Static and cyclic direct shear tests on carbonate sands. Geotechnical

Testing Journal, 1992, 15(2): 138-157.

Arroyo M, Butlanska J, Gens A, et al. Cone penetration tests in a virtual calibration chamber. Geotechnique, 2011, 61(6): 525-531.

Bhattacharya S, Lombardi D, Wood D M. Similitude relationships for physical modelling of monopile-supported offshore wind turbines. International Journal of Physical Modelling in Geotechnics, 2011, 11(2): 58-68.

Burland J. Shaft friction of piles in clay--a simple fundamental approach. Ground Engineering, 1973, 6(3): 30-42.

Byrne B, Leblanc C, Houlsby G. Response of stiff piles to long term cyclic loading. Geotechnique, 2010, 60(2): 79-90.

Chandler R J. The shaft friction of piles in cohesive soils in terms of effective stress. Civil Eng & Public Works Review /UK/, 1968, 60(708): 48-51.

Chaudhary M T A. FEM modelling of a large piled raft for settlement control in weak rock. Engineering Structures, 2007, 29(11): 2901-2907.

Choi Y, O'Neill M W. Soil plugging and relaxation in pipe pile during earthquake motion. Journal of geotechnical and geoenvironmental engineering, 1997, 123(10): 975-982.

Huang M, Liu Y. Axial capacity degradation of single piles in soft clay under cyclic loading. Soils and Foundations, 2015, 55(2): 315-328.

Igoe D, Doherty P, Gavin K. The Development and Testing of an Instrumented Open-Ended Model Pile. Geotechnical Testing Journal, 2010, 33(1): 72-82.

Iskander M. Behavior of Pipe Piles in Sand: Plugging & Pore-Water Pressure Generation During Installation and Loading. New York: Springer, 2010.

Janbu N. Static bearing capacity of friction piles. Sechste Europaeische Konferenz Fuer Bodenmechanik Und Grundbau, 1, 1976.

Jeong S, Ko J, Won J, et al. Bearing capacity analysis of open-ended piles considering the degree of soil plugging. Soils and Foundations, 2015, 55(5): 1001-1014.

Kraft Jr L M, Cox W R, Verner E A. Pile load tests: Cyclic loads and varying load rates. Journal of Geotechnical and Geoenvironmental Engineering, 1981, 107(GT1): 1-19.

Liu H L, Fei K, Zhou Y D, et al. Numerical analysis of internal friction of cast-in-place concrete thin-walled pipe pil. Rock and Soil Mechanics, 2004, (S2): 211-216.

Liu J W, Duan N, Cui L, et al. DEM investigation of installation responses of jacked open-ended piles. Acta Geotechnica, 2019, 14(6): 1805-1819.

Mohammad A, Sadek Y C, Jude L. Simulating shear behavior of a sandy soil under different soil conditions. Journal of Terramechanics, 2011, 48(6): 451-458.

Muqtadir A, Desai C S. Three-dimensional analysis of a pile-group foundation. International Journal for Numerical and Analytical Methods in Geomechanics, 1986, 10: 41.

Ottaviani M. Three-dimensional finite element analysis of vertically loaded pile groups. Géotechnique, 1975, 25(2): 159-174.

Paik K, Salgado R, Lee J, et al. Behavior of Open- and Closed-Ended Piles Driven Into Sands., 2003, 129(4): 296-306.

Paik K, Salgado R. Determination of bearing capacity of open-ended piles in sand. Journal of Geotechnical and Geoenvironmental Engineering, 2003, 129(1): 46-57.

Poulos H G. Cyclic Axial Loading Analysis of Piles in Sand. ASCE, 1989, 115(6): 836-852.

Poulos H G. Pile behavior-theory and application. Geotechnique, 1989, 39(3): 365-415.

Pressley J S, Poulos H G. Finite element analysis of mechanisms of pile group behavior. International Journal for Numerical and Analytical Methods in Geomechanics, 1986, 10: 213-221.

Qu L M, Ding X M, Wu C R, et al. Effects of topography on dynamic responses of single piles under vertical cyclic loading. Journal of Mountain Science, 2020, 17(1): 230-243.

Seed H B, Reese L C. The action of soft clay along friction piles. American Society of Civil Engineers Transactions, 1955, 81(12): 1-28.

Skempton A W. Cast in-situ bored piles in London clay. Geotechnique, 1959, 9(4): 153-173.

Technical specification for building pile foundation (JGJ 94—2008). Construction Science and Technology, 2012, (Z1): 38-39.

Trochanis A M, Bielak J, Christiano P. Three-dimensional Nonlinear Study of Piles. Journal of Geotechnical Engineering, 1991, 117(3): 429-447.

Tsuha C H C, Foray P Y, Jardine R J, et al. Behaviour of displacement piles in sand under cyclic axial loading. Soils and Foundations, 2012, 52(3): 393-410.